# Biological Activities of Plant Food Components: Implications in Human Health

# Biological Activities of Plant Food Components: Implications in Human Health

Editor

**Carla Gentile**

MDPI • Basel • Beijing • Wuhan • Barcelona • Belgrade • Manchester • Tokyo • Cluj • Tianjin

*Editor*
Carla Gentile
University of Palermo
Italy

*Editorial Office*
MDPI
St. Alban-Anlage 66
4052 Basel, Switzerland

This is a reprint of articles from the Special Issue published online in the open access journal *Foods* (ISSN 2304-8158) (available at: https://www.mdpi.com/journal/foods/special_issues/Biological_Activities_Plant_Food_Components_Implications_Human_Health).

For citation purposes, cite each article independently as indicated on the article page online and as indicated below:

LastName, A.A.; LastName, B.B.; LastName, C.C. Article Title. *Journal Name* **Year**, *Volume Number*, Page Range.

ISBN 978-3-0365-2530-3 (Hbk)
ISBN 978-3-0365-2531-0 (PDF)

Cover image courtesy of Carla Gentile

# Contents

# About the Editor

**Carla Gentile** obtained her master's degree in Chemistry and Pharmaceutical Technology and her PhD in Medicinal Chemistry at the University of Palermo. During her PhD, she was a visiting researcher at the Respiratory and Molecular Biology Division of the School of Medicine, University of Southampton (UK) and at the INSERM, Villejuif (France), where she was also a post-doctoral fellow (French Hematology Society, SFH). After a post-doctoral fellowship (Italian Ministry of University and Research, MIUR), in 2008, she became an assistant professor at the University of Palermo. Currently, she is an associate professor in Biochemistry at the Department of Biological, Chemical, and Pharmaceutical Sciences and Technologies, University of Palermo. Her principal scientific interest lies in the evaluation of the biological properties of natural antioxidants, in particular dietary phytochemicals. She is the co-author of more than 70 ISI-indexed publications that have received a total of more than 2000 citations.

*Editorial*

# Biological Activities of Plant Food Components: Implications in Human Health

**Carla Gentile**

Department of Biological, Chemical and Pharmaceutical Sciences and Technologies (STEBICEF), University of Palermo, Viale delle Scienze, 90128 Palermo, Italy; carla.gentile@unipa.it; Tel.: +39-091-23897423

**Citation:** Gentile, C. Biological Activities of Plant Food Components: Implications in Human Health. *Foods* **2021**, *10*, 456. https://doi.org/10.3390/foods10020456

Received: 29 January 2021
Accepted: 10 February 2021
Published: 19 February 2021

**Publisher's Note:** MDPI stays neutral with regard to jurisdictional claims in published maps and institutional affiliations.

Scientific data and epidemiological evidence collected over the last fifty years have shown that nutrition plays a decisive role in human health. While before it was considered essentially a non-disease condition, health is nowadays felt as a state of complete physical, mental, and social wellness, according to the definition of the World Health Organization. Eating properly is not only necessary to meet energy needs, avoiding syndromes associated with nutritional deficiency and excess, but it actively contributes to improve human health. The functional meaning of nutrition as surely protective and possibly therapeutic is today amply demonstrated by scientific evidence. Additionally, the increasing sensitivity to the healthy role of nutrition makes consumers more and more careful when choosing high quality foods.

The functional role of nutrition is due to specific small molecules with biological activity. These dietary compounds neither act as energy substrates or plastic materials for cells, nor as enzymatic cofactors. However, due to their peculiar bioactivity, they can benefit human health.

Plants are the most important source of bioactive molecules and still represent the main resource in the quest for new drugs. The interest in dietary phytochemicals is justified by the results from numerous epidemiological studies that demonstrate how diets rich in plant foods are able to prevent several human diseases, including cardiovascular, neoplastic, neurodegenerative, and metabolic pathologies. Of course, it is not possible to demonstrate that the positive effects resulting from the intake of a specific plant food are due to a particular component, for the presence of other phytochemicals and any synergistic effects cannot be neglected. However, experimental results showing biological activity for many pure dietary phytochemicals suggest that particular components may participate in the protective effects associated with the consumption of the plant food it can be sourced from.

The documented biological activity of phytochemicals is expressed through several protective effects, such as antioxidant, anti-inflammatory, antimicrobial, antitumor, immunomodulatory, neuroprotective, antihypertensive, antidiabetic actions. The commercial success of many plant-based supplements is based on these proven biological effects.

Although the protective effect associated with the consumption of plant foods is ascribable to a number of mechanisms, the biological activity of these phytocomponents has frequently been related to their ability to function as antioxidants. Many phytochemicals are in fact redox-active molecules and thus are able to influence the cellular redox balance. For this reason, besides protecting cells from oxidative stress phenomena, phytochemicals influence numerous redox-sensitive biological targets that regulate several important cellular functions. Additionally, the activity of phytochemicals can also involve interactions of these small molecules with various biological targets, including proteins, DNA and lipids, and consequently a possible alteration of their function [1].

This Special Issue addresses the biological activity of dietary phytochemicals, either purified [2] or in extracts from plant foods, and speculates on their potential effects on human health. The studied plant foods include edible parts (fruits, seeds, leaves, and flowers) of plants common in the Mediterranean area, such as pomegranate (*Punica granatum*) [3], oregano (*Origanum vulgare*) [4], and purslane (*Portulaca olearacea*) [5], but also of

1

several tropical plants, including annona (*Annona cherymola*) [6], herba mate (*Ilex paraguariensis*), guaranà (*Paullinia cupana*) [7], and ash gourd (*Benincasa hispida*) [8]. Bioactivity was studied in terms of radical scavenging and antioxidant [5–7,9], antimicrobial [10], and antiproliferative activity [4].

The collected results suggest that the intake of the aforementioned plant foods in the context of a balanced diet could be beneficial to health. In addition, considering the high concentration of bioactive molecules in some of the observed plant matrices, the presented results hint at the possibility of using some of the studied plant extracts in food functionalization or in the formulation of dietary supplements.

Due to their ability to exert several biological effects that are potentially useful for human health, over the years, dietary phytochemicals have drawn increasing interest in human nutrition research.

The papers collected in this Special Issue contribute to the growth of this research area.

I would like to thank all the authors and the reviewers of the papers published in this Special Issue for their great contributions and effort. I am also grateful to the editorial board members and to the staff of the journal for their kind support during the preparation of this Special Issue.

**Conflicts of Interest:** The author declares no conflict of interest.


## References

1. Caradonna, F.; Consiglio, O.; Luparello, C.; Gentile, C. Science and Healthy Meals in the World: Nutritional Epigenomics and Nutrigenetics of the Mediterranean Diet. *Nutrients* **2020**, *12*, 1748. [CrossRef]
2. Meng, X.; Zhou, J.; Zhao, C.-N.; Gan, R.-Y.; Li, H.-B. Health benefits and molecular mechanisms of resveratrol: A narrative review. *Foods* **2020**, *9*, 340. [CrossRef]
3. Di Stefano, V.; Scandurra, S.; Pagliaro, A.; Di Martino, V.; Melilli, M.G. Effect of Sunlight Exposure on Anthocyanin and Non-Anthocyanin Phenolic Levels in Pomegranate Juices by High Resolution Mass Spectrometry Approach. *Foods* **2020**, *9*, 1161. [CrossRef]
4. Nanni, V.; Di Marco, G.; Sacchetti, G.; Canini, A.; Gismondi, A. Oregano Phytocomplex Induces Programmed Cell Death in Melanoma Lines via Mitochondria and DNA Damage. *Foods* **2020**, *9*, 1486. [CrossRef]
5. Melilli, M.G.; Di Stefano, V.; Sciacca, F.; Pagliaro, A.; Bognanni, R.; Scandurra, S.; Virzì, N.; Gentile, C.; Palumbo, M. Improvement of fatty acid profile in durum wheat breads supplemented with portulaca oleracea L. quality traits of purslane-fortified bread. *Foods* **2020**, *9*, 764. [CrossRef]
6. Gentile, C.; Mannino, G.; Palazzolo, E.; Gianguzzi, G.; Perrone, A.; Serio, G.; Farina, V. Pomological, Sensorial, Nutritional and Nutraceutical Profile of Seven Cultivars of Cherimoya (Annona cherimola Mill). *Foods* **2021**, *10*, 35. [CrossRef]
7. Dabulici, C.M.; Sârbu, I.; Vamanu, E. The bioactive potential of functional products and bioavailability of phenolic compounds. *Foods* **2020**, *9*, 953. [CrossRef] [PubMed]
8. Sultana, R.; Alashi, A.M.; Islam, K.; Saifullah, M.; Haque, C.E.; Aluko, R.E. Inhibitory Activities of Polyphenolic Extracts of Bangladeshi Vegetables against $\alpha$-Amylase, $\alpha$-Glucosidase, Pancreatic Lipase, Renin, and Angiotensin-Converting Enzyme. *Foods* **2020**, *9*, 844. [CrossRef] [PubMed]
9. Taniguchi, A.; Kyogoku, N.; Kimura, H.; Kondo, T.; Nagao, K.; Kobayashi, R. Antioxidant Capacity of Tempura Deep-Fried Products Prepared Using Barley, Buckwheat, and Job's Tears Flours. *Foods* **2020**, *9*, 1246. [CrossRef] [PubMed]
10. Tamfu, A.N.; Ceylan, O.; Kucukaydin, S.; Ozturk, M.; Duru, M.E.; Dinica, R.M. Antibiofilm and Enzyme Inhibitory Potentials of Two Annonaceous Food Spices, African Pepper (Xylopia aethiopica) and African Nutmeg (Monodora myristica). *Foods* **2020**, *9*, 1768. [CrossRef] [PubMed]

 *foods*

*Article*

# Pomological, Sensorial, Nutritional and Nutraceutical Profile of Seven Cultivars of Cherimoya (*Annona cherimola* Mill)

Carla Gentile [1,*], Giuseppe Mannino [2], Eristanna Palazzolo [3], Giuseppe Gianguzzi [3], Anna Perrone [1], Graziella Serio [1] and Vittorio Farina [3]

[1]  Department of Biological, Chemical and Pharmaceutical Sciences and Technologies (STEBICEF), University of Palermo, Viale delle Scienze, 90128 Palermo, Italy; anna.perrone@unipa.it (A.P.); graziella.serio01@unipa.it (G.S.)
[2]  Department of Life Sciences and Systems Biology, Innovation Centre, Plant Physiology Unit, University of Turin, Via Quarello 15/A, 10135 Turin, Italy; giuseppe.mannino@unito.it
[3]  Department of Agricultural, Food and Forest Sciences (SAAF), University of Palermo, Viale delle Scienze, 90128 Palermo, Italy; eristanna.palazzolo@unipa.it (E.P.); giuseppe.gianguzzi@unipa.it (G.G.); vittorio.farina@unipa.it (V.F.)
*  Correspondence: carla.gentile@unipa.it; Tel.: +39-0912-389-7423

**Abstract:** In this work, the food quality of four international (*Campas, Chaffey, Fino de Jete* and *White*) and three local (*Daniela, Torre1* and *Torre2*) cultivars of Cherimoya (*Annona cherimola* Mill) was investigated. With this aim, pomological traits, sensorial attributes, physiochemical parameters (pH, total soluble content and total acidity), nutritional composition (macro- and micro-nutrients) and nutraceutical values (bioactive compounds, radical scavenging and antioxidant properties) were evaluated. Among the seven observed cultivars, *Fino de Jete* was identified as the best, not only for its commercial attributes such as pomological traits and physiochemical values, but also for its nutritional composition. On the other hand, *Chaffey* and *Daniela* were the cultivars with the highest content of polyphenols, proanthocyanidins, and with the strongest antioxidant capacity. Concerning the two local ecotypes, *Torre1* and *Torre2*, they displayed a balanced nutritional profile that, if combined with their discrete nutraceutical, physicochemical and pomological values, may result in a reassessment of their commercial impact. In conclusion, our data provide interesting information about the pomological, nutritional, and nutraceutical properties of cherimoya fruits. Our results, in addition to promoting the commercial impact of local cultivars, may increase the use of individual cultivars in breeding programs.

**Keywords:** polyphenols; sensory analysis; mineral content; proanthocyanidins; carotenoids; antioxidant activity; FRAP; DPPH; ABTS; CAA

**Citation:** Gentile, C.; Mannino, G.; Palazzolo, E.; Gianguzzi, G.; Perrone, A.; Serio, G.; Farina, V. Pomological, Sensorial, Nutritional and Nutraceutical Profile of Seven Cultivars of Cherimoya (*Annona cherimola* Mill). *Foods* **2021**, *10*, 35. http://dx.doi.org/10.3390/foods10010035

Received: 9 November 2020
Accepted: 21 December 2020
Published: 24 December 2020

**Publisher's Note:** MDPI stays neutral with regard to jurisdictional claims in published maps and institutional affiliations.

## 1. Introduction

The *Annonaceae* Juss. family covers more than 2000 species, of which 120 belong to the genus *Annona* L. [1]. The most famous species are *Annona cherimola* Mill (cherimoya), *Annona muricata* L. (soursop), *Annona squamosa* L. (conde fruit), *Annona reticulata* L. (custard apple), and the interspecific hybrid *Atemoya* (*A. cherimola* × *A. squamosa*). In particular, *Annona cherimola* is the most diffused specie in subtropical countries. It is an indigenous tree of Andean South America and it has naturalized in tropical highlands and subtropical areas of South America [2]. The marketable value of cherimoya is related to its big, heart-shaped and conical fruit [3], which may reach considerable weight and size. The edible flesh of these fruits is white, creamy, and with a custard-like consistency [4]. The aromatic flavor is a mix of papaya, banana, pineapple, and passion fruit [5]. Although cherimoya fruit is consumed as fresh fruit, it can be also processed making yogurt, ice creams and other desserts. It is not recommended to ripen the fruits on the tree, because they lose quality. Cherimoya fruits are generally harvested when not fully mature, and left for ripening

3

under controlled storage conditions. The time of harvesting is commonly determined by the changes of skin fruit color, which turn from green to yellowish-green in the proximity of physiological maturity [6]. In Europe, Spain is the most important producer of cherimoya fruits, and the most important cultivars are *Campas* and *Fino de Jete*, which are also the most widespread cultivars in the global market. In Italy, *A. cherimola* is well adapted to the pedoclimatic conditions of the tyrrhenian coastal areas of Sicily and Calabria, where there are good climate conditions for the production of other exotic and tropical fruits, such as loquat, mango, litchi, avocado, banana and papaya [7–10]. In particular, in Sicily, in addition to affirmed cultivars such as *Fino de Jete*, local ecotypes are also cultivated with very limited diffusion. Concerning the nutritional value, cherimoya flesh possesses a high content of sugars, while having low fat content and, in comparison with other tropical fruits, also good Ca and P content [11]. However, although the nutraceutical properties of other *Annona* species have been extensively studied, those related to cherimoya fruits are much less investigated. The majority of the available literature data suggests that this tropical fruit is an interesting source of bioactive compounds, including polyphenols (catechin, proanthocyanidins, hydroxytyrosol) [12,13], alkaloids (annocherines, norisocorydine, cheritamine, annonaine) [14,15], acetogenins (cherimolin-2 and almunequin) [16], terpenes (myrcene, pinene, linalool, caryophillene, terpenolene and germacrene) [17] and cyclopeptides (cherimola cyclopetide E and cherimola cyclopetide F) [18,19]. In addition, antioxidant [20], pro-apoptotic [3,16,17,21], anti-protozoal [22], and anti-diabetic [23] activities were also demonstrated for extracts obtained from different part of the fruit.

The aim of this study was the investigation of the pomological, physiochemical, sensorial, nutritional and nutraceutical attributes of seven cultivars of cherimoya fruits grown in Sicily. Our results provide comprehensive information on the quality of cherimoya fruits and can be useful for the improvement of the utilization of the specific genotype.

## 2. Materials and Methods

### 2.1. Plant Material

The fruits were obtained from trees grown in Vivai Torre (Milazzo, Sicily, Italy; 38°19′ N, 15°24′ E; 20 m a.s.l.). Four international affirmed CVs (*Campas, Chaffey, Fino de Jete* and *White*) and three local CVs (*Daniela, Torre1* and *Torre2*) were selected (Table 1). The fruits of each CV were picked from three 15-year-old trees, grafted on their own rootstock. The trees were planted in North-South direction, with an inter-tree spacing of 6 m and 6 m between rows. The yield per tree was measured by weighing and counting the total number of fruits per tree, and at each harvest time, the trunk circumference was measured at ~15 cm above the graft union. The yield efficiency and crop load were expressed in kilograms or number of fruits per trunk of a cross-sectional area (TCSA) or leaf area. A sample of 30 fruits per CV (10 fruits per 3 trees) was hand-picked when not fully mature, and the color changed from green to yellow. After harvest, the fruits were left to ripen under storage conditions (20 °C). Fifteen fruits were employed for pomological and physicochemical analysis, and 5 of them were employed for the sensorial evaluation. Finally, the others 15 fruits were immediately frozen in liquid nitrogen and then stored at −80 °C until the analysis of nutrients, bioactive compounds, and antioxidant activity was performed within 3 months. Immediately before the analysis, three fruits for each CV were thawed and peeled, and then the seeds were removed. The chopped pulp was homogenated. Three aliquots of each homogenate for each analysis were employed.

**Table 1.** The name of cultivar (CV), origin, yield, number of fruits per tree, yield efficiency percentage, and crop load of the observed seven *Annona cherimola* cultivars

| CV | Origin | Harvest Date | Tree Vigour | Yield (kg/tree) | Fruits Per Tree | %Yield Efficiency (kg/cm$^2$) | %Crop Load (fruit/cm$^2$) |
|---|---|---|---|---|---|---|---|
| *Campas* | Spain | 08 Nov | High | $7.25 \pm 0.63$ [e] | $50.5 \pm 3.0$ [cb] | $1.01 \pm 0.05$ [e] | $8.12 \pm 0.51$ [e] |
| *Chaffey* | USA-California | 03 Dec | High | $14.04 \pm 0.64$ [c] | $65.0 \pm 3.0$ [b] | $4.23 \pm 0.19$ [c] | $19.55 \pm 0.47$ [a] |
| *Daniela* | Italy | 02 Dec | Mdium-low | $30.11 \pm 0.12$ [a] | $80.5 \pm 2.0$ [a] | $4.21 \pm 0.21$ [c] | $12.14 \pm 0.43$ [c] |
| *Fino de Jete* | Spain | 02 Dec | Medium | $30.41 \pm 0.78$ [a] | $80.0 \pm 3.0$ [a] | $6.17 \pm 0.26$ [a] | $16.77 \pm 0.21$ [b] |
| *Torre1* | Italy | 07 Dec | Medium | $17.15 \pm 0.29$ [b] | $45.0 \pm 3.0$ [cd] | $4.55 \pm 0.19$ [c] | $11.23 \pm 0.41$ [cd] |
| *Torre2* | Italy | 08 Nov | High | $8.84 \pm 0.54$ [d] | $26.0 \pm 2.0$ [e] | $5.23 \pm 0.37$ [b] | $15.41 \pm 0.42$ [c] |
| *White* | USA-California | 03 Dec | High | $6.83 \pm 0.45$ [f] | $35.5 \pm 2.0$ [d] | $2.41 \pm 3.21$ [d] | $10.11 \pm 0.41$ [d] |

Values are expressed as mean $\pm$ SD of data collected in two years. Among the same series, different lowercase letters indicate significantly different values at $p \leq 0.05$, as measured by Tukey's tests.

## 2.2. Pomological and Physiochemical Analysis

Fruit weight (FW), longitudinal diameter (LD), transversal diameter (TD), seed weight (SW), peel weight (PeW), pulp weight (PW), flesh firmness (FF), total soluble solids content (TSSC), and titratable acidity (TA) were measured. FW, SW, PeW and PW (g) were determined using a digital scale (Gibertini EU-C 2002 RS, Novate Milanese, Italy); LD and TD (mm) using a digital calliper TR53307 (Turoni, Forlì, Italy); FF (kg/cm$^2$) using a digital penetrometer TR5325 with a 8 mm diameter tip (Turoni, Forlì, Italy); TSSC ($^\circ$Brix) using a digital refractometer Atago Palette PR-32 (Atago Co., Ltd., Tokyo, Japan) and TA (g citric acid per L) using a CrisonS compact tritator (Crison Instruments, SA, Barcelona, Spain). Skin and flesh colors were calculated using a Konica Minolta Colorimeter based on the CIELAB system that measured the lightness (L*) and the variation from red (+a*) to green ($-$a*), and from green (+b*) to yellow ($-$b*) in the fruits.

## 2.3. Sensory Analyses

A trained panel consisting of 10 judges (5 females and 5 males, between 22 and 45 years of age) performed the sensory profile analysis, as previously reported [7]. All panelists were trained and developed wide expertise in sensory evaluation of tropical fruits. The judges in preliminary sessions generated 24 sensory descriptors (Table 2), and they evaluated samples using a hedonic scale, assigning to each descriptor a score from 1 (absence of the sensation) to 9 (highest intensity). The order of each sample was randomized for each panelist, and water was provided for rinsing between the different samples.

**Table 2.** A list of the evaluated sensory descriptors and their definitions.

| Descriptors | Acronyms | Definition |
|---|---|---|
| **Appearance** | | |
| Skin Color | SC | Predominant color of the main surface of the Cherimoya |
| Flesh Color | FC | Color of the Cherimoya flesh (from pale green to dark green) |
| **Aroma** | | |
| Off-Odor | OFO | Non characteristic odor |
| Exotic Fruit Odor | EXO | Characteristic aroma of exotic fruit perceived with the sense of smell |
| Melon, Banana and Pear Odor | MBO | *Annona* characteristic odor |
| Medicine Odor | MO | Non characteristic odor |
| Grassy Odor | GO | Characteristic aroma of cut grass perceived with the sense of smell |

**Table 2.** *Cont.*

| Descriptors | Acronyms | Definition |
|---|---|---|
| Fruity Odor | FO | Fruit characteristic aroma |
| **Flavor** | | |
| Melon, Banana and Pear Flavor | MBF | *Annona* characteristic odor |
| Fermented Flavor | FFL | Characteristic flavor of fruit at initial fermentation process |
| Exotic Fruit Flavor | EXF | Exotic fruit characteristic flavor |
| Fruity Flavor | FFL | Fruit characteristic flavor |
| Flavor Alcohol | ALF | Flavor associated with alcohol scent |
| Off-flavor | OFF | Non characteristic odor |
| **Taste and tactile in mouth** | | |
| Acid | A | Basic taste on tongue stimulated by acids |
| Astringent | AST | Sensory perception in the oral cavity that may include drying sensation, and roughing of the oral tissue |
| Sweetness | S | Taste on the tongue stimulated by sugars and high potency sweeteners |
| Pungent | P | Sensation of tingling perceived in the oral cavity |
| **Rheological** | | |
| Juiciness | J | The amount of juice/moisture perceived in the mouth. |
| Consistency | C | The force it takes to bite through the sample |
| General Appearance | APP | Regularity of shape, size, gloss, color and absence of defects |
| Mellowness | M | Perceived time during swallowing |
| Overall Evaluation | OVE | Overall judgment |
| Mealiness | MEA | A flour-like texture |

*2.4. Nutritional Parameters*

2.4.1. Content of Carbohydrates, Lipids, Proteins, Water and Ashes

The carbohydrate and protein content were evaluated as previously described, respectively using Anthrone's [24] and Kjedahl's [25] methods. Ash and water contents were determined through the procedure described in AOAC [26]. The content of lipids was calculated after lipid extraction with a gravimetric method, as previously described [27]. Data were expressed as g per 100 g of Pulp Weight (PW).

2.4.2. Mineral Content

The contents of K, Na, Ca, Mg, Fe, Cu, Mn, and Zn were determined by atomic absorption spectroscopy following wet mineralization, and using the instrumental condition as previously described [28]. Briefly, the samples were digested, and approximately 100 mg of dried sample was weighed and incubated with 9 mL of 65% ($w/w$) $HNO_3$, and 1 mL of 30% ($w/v$) $H_2O_2$ were added. The temperature was set at 200 °C for 20 min. Once cooled, the digested samples were diluted to a final volume of 50 mL with distilled $H_2O$. All measurements were performed using an Agilent 4200 MP-AES fitted with a double-pass cyclonic spray chamber and OneNeb nebulizer. The calibration standards were prepared by diluting a 1000 mg/L multi-element standard solution (Sigma Aldrich and Scharlab S.L.) in 1% ($v/v$) $HNO_3$. Finally, P was determined using a colorimetric method [29]. Data were expressed as mg per 100 g of Fresh Weight (PW).

2.4.3. Vitamin Content

Retinol (Vit. A), Riboflavin (Vit. B2), Niacin (Vit. B3), and Ascorbic Acid (Vit. C) were extracted and determined according to previously reported methods. Briefly, Vit. A was extracted and quantified using a commercial kit (Vitamin A Food ELISA Kit, Crystal Chem, NL) and following the manufacturer's instructions. Vit. B1 and Vit. B2 were respectively extracted using 0.1 N HCl [30] or a solution of 1% ($v/v$) $H_2SO_4$ [31]. Quantification was performed via HPLC equipped with a fluorimetric detector [30,31]. Finally, Vit C was extracted with 10 mL of 1% ($v/v$) $HPO_3$ for 45 min from dried extract, previously prepared [7]. After filtration, 1 mL was mixed with 9 mL of $C_{12}H_7NC_{12}O_2$ and the absorbance was measured at 515 nm against a blank after 30 min. Vitamin C was quantified using a calibration curve of authentic L-ascorbic acid (0.02–0.12 mg/100 g). Data were expressed as mg per 100 g of Fresh Weight (PW).

*2.5. Bioactive Compounds and Antioxidant Activities*

2.5.1. Total Carotenoid Content (TCC)

TCC was determined in flesh homogenates via spectrophotometric analysis after extraction of carotenoids, as previously reported [8]. Data were expressed as β-carotene Equivalent per 100g of PW, using the molecular weight (536.87 g mol$^{-1}$) and the molar extinction coefficient (2505 M$^{-1}$ cm$^{-1}$) of β-carotene in Hexane.

2.5.2. Preparation of Fruit Extracts

Three aliquots of each homogenate were extracted twice with 70% (*v/v*) EtOH using a 1:20 (*w/v*) ratio. After centrifugation (10 min at 10.000 *g*, 4 °C) and filtration through a Millex HV 0.45 μm filter (Millipore, Billerica, MA), the supernatants were recovered and combined together. Ethanolic extracts were used both for the determination of bioactive compounds and the antioxidant properties.

2.5.3. Total Polyphenol Content (TPC)

The phenolic content of the flesh of the observed CVs of *A. cherimola* was determined in ethanolic extracts via the Folin-Ciocalteu method, with some minor changes as previously reported [32,33]. Results were expressed as mg Gallic Acid Equivalents (GAE) per 100 g of PW.

2.5.4. Determination of the Total ProAnthocyanidins Content (TPAC) and Investigation of the Polymerization Linkage via HPLC-MS/MS

The proanthocyanidins (PACs) were evaluated in the ethanolic extracts via BL-DMAC assay [34] with some minor changes, as previously reported [35]. The PAC concentration in the extracts was expressed as mg PAC-A equivalent per 100 g of PW.

In order to investigate PAC grade and type, polymerization-binding of catechins was investigated via High-Pressure Liquid Chromatography (HPLC, Agilent 1260, Technologies, Santa Clara, CA, USA) coupled with 6330 Series Ion Trap (Agilent Technologies, USA), as previously reported [9].

2.5.5. Radical Scavenging and Metal Ion Reducing Activity

The antioxidant activity of the ethanolic extracts was measured evaluating both the radical scavenging activity via ABTS [36] and DPPH [37], and the reducing antioxidant power via FRAP [38] assays. Data were expressed as mmol Trolox Equivalent (TE) per 100 of PW as previously reported [35].

2.5.6. Cellular Antioxidant Activity Assay (CAA)

The CAA assay was performed as previously described by Wolfe at al [39], with some minor changes [40]. For the experiments, we used HepG2 (human liver cancer cell line), obtained from American Type Culture Collection (ATCC) (Rockville, MD, USA). The antioxidant activity was expressed as CAA$_{50}$ that is the amount of flesh in cell medium necessary to obtain the 50% of inhibition of oxidative stress, with respect to the positive control. CAA$_{50}$ was calculated from concentration-response curves using linear regression analysis, and it was expressed as μg of PW per mL of cell medium.

*2.6. Statistical Analysis*

Each assay was repeated three times. All data were tested for differences between the CVs using one-way analysis of variance (ANOVA; general linear model) followed by Tukey's multiple range test for $p \leq 0.05$, marking significant differences among the samples with different lowercase letters. Principal Component Analysis (PCA) and HeatMap Cluster Analysis were performed using covariant matrix of extraction and varimax rotation.

All statistical analyses were performed using SPSS ver. 24. The nucleotide sequences were analysed via CLC software, and the cladogram of gene sequences was performed with ClustalX software by using the Neighbour Joining (NJ) method. Bootstrap values were calculated from 100 resampling of the alignment data.

### 3. Results and Discussion

*3.1. Pomological and Physiochemical Parameters*

The fruits of the observed *A. cherimola* CVs showed wide variability of the pomological (Table 3) and physiochemical (Table 4) parameters. All the observed CVs, except *Campas* and *White*, reached a considerable size, with a small incidence of SW and PeW on the total FW. For these CVs, the edible part ranged between 73% (*Daniela*) and 87% (*Fino de Jete*) of the total FW. Moreover, *Fino de Jete* and *Daniela* produced the largest and biggest fruits. On the other hand, *Campas* and *White* produced the smallest fruits, with an incidence of non-edible part of about 40%. In addition, significant differences in FF among the different CVs were not recorded. Concerning the edible part, generally, the largest fruits showed also the highest percentage of flesh (PW/FW). The highest yield per tree was obtained in *Daniela* and *Fino de* Jete, whereas *White*, *Campas* and *Torre 2* showed very low values (Table 1). Yield improvement was caused by the increase of the fruit size, rather than to the number of fruits; nevertheless, crop load was higher for *Fino de Jete* and *Chaffey*. Moreover, the highest yield efficiency was observed in *Chaffey* followed by *Fino de Jete*.

**Table 3.** The pomological traits of the seven observed *Annona cherimola* fruits. Data are expressed as mean $\pm$ SD. For each row, different lowercase letters mark significant ($p < 0.05$) differences among the samples, as measured by one-way ANOVA followed by Tuckey's test. The letter "a" denotes the highest content.

|  | *Campas* | *Chaffey* | *Daniela* | *Fino de Jete* | *Torre1* | *Torre2* | *White* |
|---|---|---|---|---|---|---|---|
| FW | 280 $\pm$ 26.0 [c] | 392 $\pm$ 182 [bc] | 578 $\pm$ 163 [a] | 607 $\pm$ 121 [a] | 454 $\pm$ 62.6 [b] | 485 $\pm$ 71.1 [ab] | 362 $\pm$ 110 [bc] |
| PW | 158 $\pm$ 80.4 [c] | 290 $\pm$ 90.6 [b] | 423 $\pm$ 83.5 [ab] | 527 $\pm$ 85.5 [a] | 341 $\pm$ 90.3 [b] | 375 $\pm$ 90.8 [b] | 229 $\pm$ 45.7 [bc] |
| PeW | 49.8 $\pm$ 8.21 [c] | 74.7 $\pm$ 24.3 [ab] | 92.7 $\pm$ 19.9 [a] | 93.3 $\pm$ 6.70 [a] | 80.4 $\pm$ 6.80 [ab] | 70.4 $\pm$ 8.10 [b] | 59.9 $\pm$ 12.3 [bc] |
| SW | 16.2 $\pm$ 4.42 [ab] | 15.6 $\pm$ 10.0 [ab] | 23.5 $\pm$ 6.12 [ab] | 25.6 $\pm$ 5.81 [a] | 14.0 $\pm$ 7.00 [ab] | 14.2 $\pm$ 6.00 [ab] | 13.4 $\pm$ 3.71 [b] |
| SN | 31.6 $\pm$ 9.40 [ab] | 22.8 $\pm$ 14.3 [b] | 45.0 $\pm$ 6.81 [a] | 44.5 $\pm$ 17.7 [a] | 24.0 $\pm$ 11.1 [ab] | 34.3 $\pm$ 7.80 [ab] | 25.3 $\pm$ 5.12 [ab] |
| LD | 85.9 $\pm$ 11.3 [c] | 78.3 $\pm$ 9.70 [c] | 102 $\pm$ 11.9 [b] | 120 $\pm$ 15.5 [a] | 119 $\pm$ 10.7 [ab] | 115 $\pm$ 16.0 [ab] | 77.5 $\pm$ 9.30 [c] |
| TD | 80.3 $\pm$ 4.20 [b] | 103 $\pm$ 9.61 [a] | 98.4 $\pm$ 6.90 [a] | 98.9 $\pm$ 6.10 [a] | 77.8 $\pm$ 5.00 [b] | 93.5 $\pm$ 7.61 [a] | 92.8 $\pm$ 12.1 [a] |
| FF | 1.01 $\pm$ 0.60 [ab] | 1.90 $\pm$ 1.51 [a] | 1.72 $\pm$ 0.40 [ab] | 0.81 $\pm$ 0.90 [b] | 0.60 $\pm$ 0.51 [b] | 1.60 $\pm$ 0.41 [ab] | 1.01 $\pm$ 1.11 [ab] |

FW is the Fruit Weight; PW is the Pulp Weight; PeW is the Peel Weight; SW is the Seed Weight; SN is the Seed Number; LD is the Longitudinal diameter; TD is the Transversal diameter; F is the firmness.

The L*, a* and b* parameters of peel and pulp of Cherimoya fruits were minimally influenced by the genotype (Table 4). Low a* and high b* values were recorded in both peel and pulp for all the observed CVs, indicating a brown peel color and a yellow color of the pulp. It was previously suggested that during the maturation of *A. crassiflora* fruits, the decrease of a* and the increasing of b* may be related to chlorophyll degradation and carotenoid accumulation, typical of ripening processes [41].

Also, TSSC varied low among the analyzed CVs, and the mean value recorded was 19.4 $\pm$ 1.82 °Brix, and the highest values were recorded in *Chaffey* and *Torre1* (Table 4). Our results were similar to those of Andrès-Augustin and colleagues who recorded comparable ranges for TSSC in fruits from commercial and local CVs of Cherimoya from Mexico [42]. Concerning TA, the observed CVs may be grouped in two different subgroups: the first one included *Fino de Jete*, *Torre1*, *Torre2* and *White*, showing a TA value more than 4.0 g malic acid per L; the second one included *Campas*, *Chaffey*, and *Daniela*, with a TA value less than 4.0 g of malic acid per L (Table 4). When the TSSC/TA ratio is considered, the fruits of the observed CVs may be divided between the sweetest (*Campas*, *Cheffey* and *Daniela*) and the bitterest (*Fino de Jete* and *White*). On the other hand, *Torre1* and *Torre2*, the two local CVs, displayed intermediate behaviors.

**Table 4.** Color and physiochemical parameters of the seven observed *Annona cherimola* fruits. For each parameter, different letters indicate significant ($p < 0.05$) differences among the cultivars as measured by one-way ANOVA followed by Tuckey's test. Letter "a" denotes the highest content.

| | Colour Parameters | | | | | |
| --- | --- | --- | --- | --- | --- | --- |
| | Peel Colour of Fruits | | | Pulp Colour of Fruits | | |
| | L* | a* | b* | L* | a* | b* |
| *Campas* | 61.1 ± 4.61 [bc] | −11.9 ± 1.71 [c] | 30.0 ± 2.82 [a] | 78.4 ± 3.81 [a] | −1.06 ± 0.85 [ab] | 15.9 ± 2.02 [a] |
| *Chaffey* | 57.6 ± 4.82 [c] | −8.88 ± 1.82 [ab] | 29.5 ± 2.92 [a] | 82.8 ± 2.91 [a] | −0.92 ± 0.31 [ab] | 13.7 ± 1.83 [ab] |
| *Daniela* | 59.1 ± 3.71 [bc] | −11.0 ± 2.21 [b] | 29.5 ± 2.33 [a] | 81.4 ± 5.69 [a] | −1.06 ± 0.54 [ab] | 14.6 ± 2.02 [ab] |
| *Fino de Jete* | 69.6 ± 2.72 [a] | −8.73 ± 1.62 [a] | 30.6 ± 2.64 [a] | 83.5 ± 2.78 [a] | −1.14 ± 0.22 [ab] | 13.0 ± 1.74 [b] |
| *Torre1* | 63.2 ± 4.01 [b] | −11.5 ± 1.93 [bc] | 31.1 ± 3.19 [a] | 82.0 ± 2.69 [a] | −1.53 ± 0.25 [b] | 12.0 ± 1.88 [b] |
| *Torre2* | 63.1 ± 1.48 [b] | −10.3 ± 2.55 [b] | 30.2 ± 1.97 [a] | 82.3 ± 3.39 [a] | −0.91 ± 0.24 [ab] | 13.9 ± 1.59 [ab] |
| *White* | 59.7 ± 2.90 [bc] | −10.5 ± 1.41 [b] | 30.4 ± 1.02 [a] | 81.3 ± 3.91 [a] | −0.81 ± 0.24 [a] | 13.3 ± 1.58 [b] |

| | Physiochemical Parameters | | | |
| --- | --- | --- | --- | --- |
| | TSSC | TA | TSSC/TA | pH |
| | (°Brix) | (g L$^{-1}$ of malic acid) | | |
| *Campas* | 18.9 ± 1.01 [b] | 3.57 ± 0.81 [c] | 7.56 ± 7.40 [a] | 5.23 ± 0.91 [bc] |
| *Chaffey* | 22.2 ± 4.02 [a] | 3.69 ± 0.70 [c] | 6.67 ± 2.61 [a] | 4.64 ± 0.33 [c] |
| *Daniela* | 19.1 ± 1.32 [b] | 3.76 ± 0.40 [c] | 6.08 ± 3.40 [a] | 5.75 ± 0.44 [ab] |
| *Fino de Jete* | 16.9 ± 0.81 [b] | 4.61 ± 0.80 [ab] | 4.48 ± 2.50 [b] | 5.23 ± 0.45 [bc] |
| *Torre1* | 21.6 ± 1.33 [a] | 4.64 ± 0.32 [ab] | 4.91 ± 1.50 [ab] | 5.01 ± 0.31 [bc] |
| *Torre2* | 19.0 ± 0.84 [b] | 4.43 ± 0.71 [ab] | 5.18 ± 2.81 [ab] | 6.02 ± 1.42 [a] |
| *White* | 18.7 ± 1.01 [b] | 5.26 ± 0.40 [a] | 4.18 ± 2.51 [b] | 4.90 ± 0.11 [c] |

TSSC is the Total Solid Soluble Content; TA is the Titratable Acidity; TSSC/TA is the ratio between the Total Solid Soluble Content; L* is the lightness; a* is the variation from red (+a*) to green (−a*); b* is the variation from green (+b*) to yellow (−b*). Data are mean values ± SD.

### 3.2. Sensorial Analysis

The panel evaluation produced sensory profiles indicating that both commercial and local CVs have good organoleptic characteristics for fresh consumption, thanks to the good combination of some key attributes (Figure 1). In particular, the fruits recorded high values for appearance (APP), skin color (SC), flesh color (FC), consistency (C), juiciness (J), and melon, banana, and pear odor (MBO). The combination of these parameters with the low values for the medicine (MO) and grassy odor (GO) resulted in a good fruit appeal for the consumer.

Concerning the taste, sensorial analysis data showed that the observed fruits had high values of sweetness (S) and low values for astringent (AST), pungent (P), and acid (A) tastes. In particular, acidic taste is a sensorial attribute known to be one of the most important quality traits for the consumer, and its perception may be correlated with the TSSC/TA ratio. However, the lack of significant Pearson correlation between S and TSSC/TA ratio is not surprising since this correlation is typically stronger for more bitter fruits [43]. Sensorial analysis also suggested significant differences among the CVs concerning specific parameters such as juiciness (J), pulp (FC) and peel (PC) color and consistency (C). In particular, panel evaluations suggested that fruits with the lowest FF (*Fino de Jete* and *Torre1*) are those more appreciated because of their consistency.

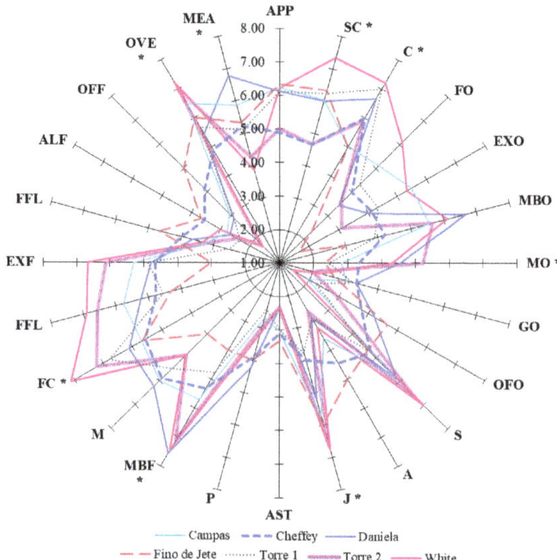


**Figure 1.** Sensorial descriptors of the observed seven cultivars of *Annona cherimola* fruits. Values are represented as mean ± SD. For each series, the symbol "*" indicate statistical ($p < 0.05$) differences among the different cultivars, as measured by Student's *t*-test.

### 3.3. Nutritional Parameters

The nutritional values per 100 g of the pulp of Cherimoya fruits are shown in Table 5. We recorded a mean moisture content equal to 79.33 ± 1.11 g per 100 g of PW. The mean value for the content of proteins, fats, and sugars was equal to 1.62 ± 0.14, 0.22 ± 0.04, and 14.09 ± 1.23 g per 100 g of PW, respectively. These values are in accordance with those reported from Morton [44]. Nerveless, strong differences in the macronutrients among the observed seven CVs were not measured, *Chaffey* was the CV with the highest sugar content, and *Fino de Jete* the CV with the lowest.

**Table 5.** The nutritional, mineral, and vitamin composition of the seven observed *Annona cherimola* fruits. Data are mean values $\pm$ SD, and they are expressed per 100 of Pulp Weight (PW). For each row, different letters mark statistical ($p < 0.05$) differences among the cultivars, as measured by one-way ANOVA analysis followed by Tuckey's test. The letter "a" denotes the highest content.

| | Campas | Chaffey | Daniela | Fino de Jete | Torre1 | Torre2 | White |
|---|---|---|---|---|---|---|---|
| **Nutritional Content** (g per 100 g of PW) | | | | | | | |
| Water Content | 78.4 $\pm$ 0.39 [ab] | 78.9 $\pm$ 0.68 [ab] | 77.5 $\pm$ 0.61 [b] | 79.3 $\pm$ 0.93 [ab] | 80.3 $\pm$ 0.63 [a] | 80.4 $\pm$ 1.26 [a] | 80.3 $\pm$ 0.91 [a] |
| Ashes | 0.70 $\pm$ 0.07 [a] | 0.68 $\pm$ 0.07 [ab] | 0.63 $\pm$ 0.07 [ab] | 0.72 $\pm$ 0.02 [a] | 0.53 $\pm$ 0.01 [b] | 0.59 $\pm$ 0.01 [ab] | 0.69 $\pm$ 0.06 [ab] |
| Protein Content | 1.75 $\pm$ 0.10 [a] | 1.75 $\pm$ 0.13 [a] | 1.69 $\pm$ 0.08 [ab] | 1.69 $\pm$ 0.18 [ab] | 1.31 $\pm$ 0.11 [b] | 1.56 $\pm$ 0.13 [ab] | 1.56 $\pm$ 0.12 [ab] |
| Fat Content | 0.19 $\pm$ 0.04 [ab] | 0.23 $\pm$ 0.01 [a] | 0.25 $\pm$ 0.01 [a] | 0.14 $\pm$ 0.01 [b] | 0.29 $\pm$ 0.02 [ab] | 0.21 $\pm$ 0.01 [a] | 0.25 $\pm$ 0.02 [a] |
| Sugar Content | 13.0 $\pm$ 1.31 [ab] | 16.0 $\pm$ 0.57 [a] | 14.7 $\pm$ 2.09 [ab] | 12.5 $\pm$ 0.89 [b] | 13.1 $\pm$ 0.53 [ab] | 14.9 $\pm$ 1.35 [ab] | 14.2 $\pm$ 0.49 [ab] |
| Raw Fibber Content | 4.32 $\pm$ 0.11 [a] | 2.35 $\pm$ 0.08 [b] | 4.41 $\pm$ 0.49 [a] | 5.35 $\pm$ 0.26 [a] | 4.61 $\pm$ 0.95 [a] | 2.11 $\pm$ 0.07 [b] | 2.93 $\pm$ 0.07 [b] |
| Energy (KJoule) | 254 $\pm$ 8.94 [c] | 306 $\pm$ 5.62 [a] | 284 $\pm$ 3.66 [ab] | 243 $\pm$ 3.53 [d] | 254 $\pm$ 8.63 [c] | 283 $\pm$ 8.01 [ab] | 274 $\pm$ 9.01 [b] |
| **Mineral Content** (mg per 100 g of PW) | | | | | | | |
| K | 273.3 $\pm$ 13.2 [ab] | 169 $\pm$ 6.50 [c] | 246 $\pm$ 25.0 [b] | 303 $\pm$ 34.6 [a] | 182 $\pm$ 4.35 [c] | 146 $\pm$ 11.3 [c] | 178 $\pm$ 3.63 [c] |
| Na | 20.0 $\pm$ 7.02 [a] | 19.3 $\pm$ 6.42 [a] | 21.6 $\pm$ 9.60 [a] | 19.3 $\pm$ 5.03 [a] | 17.3 $\pm$ 1.15 [a] | 23.3 $\pm$ 10.0 [a] | 18.3 $\pm$ 4.72 [a] |
| Ca | 10.0 $\pm$ 1.01 [a] | 9.00 $\pm$ 1.02 [a] | 9.01 $\pm$ 1.73 [a] | 9.33 $\pm$ 1.15 [a] | 9.33 $\pm$ 1.15 [a] | 8.01 $\pm$ 2.12 [a] | 9.02 $\pm$ 1.01 [a] |
| Mg | 10.0 $\pm$ 1.11 [a] | 8.66 $\pm$ 0.57 [a] | 9.66 $\pm$ 2.51 [a] | 10.3 $\pm$ 0.57 [a] | 10.1 $\pm$ 2.02 [a] | 8.02 $\pm$ 1.03 [a] | 9.33 $\pm$ 1.15 [a] |
| P | 31.6 $\pm$ 2.08 [a] | 22.5 $\pm$ 9.19 [a] | 28.0 $\pm$ 3.02 [a] | 31.0 $\pm$ 3.03 [a] | 22.6 $\pm$ 2.08 [a] | 18.2 $\pm$ 5.29 [a] | 24.3 $\pm$ 2.51 [a] |
| S | 20.6 $\pm$ 2.08 [a] | 25.3 $\pm$ 4.93 [a] | 19.3 $\pm$ 4.72 [a] | 24.3 $\pm$ 2.08 [a] | 19.6 $\pm$ 3.51 [a] | 20.3 $\pm$ 4.52 [a] | 22.3 $\pm$ 2.08 [a] |
| Cl | 88.0 $\pm$ 12.5 [a] | 96.0 $\pm$ 9.64 [a] | 80.0 $\pm$ 6.55 [a] | 89.6 $\pm$ 17.4 [a] | 71.6 $\pm$ 1.52 [a] | 87.3 $\pm$ 9.29 [a] | 86.6 $\pm$ 1.52 [a] |
| Cu | 0.16 $\pm$ 0.09 [a] | 0.09 $\pm$ 0.02 [a] | 0.18 $\pm$ 0.06 [a] | 0.13 $\pm$ 0.02 [a] | 0.19 $\pm$ 0.06 [a] | 0.13 $\pm$ 0.07 [a] | 0.09 $\pm$ 0.05 [a] |
| Mn | 0.11 $\pm$ 0.02 [a] | 0.12 $\pm$ 0.01 [a] | 0.10 $\pm$ 0.02 [a] | 0.08 $\pm$ 0.06 [a] | 0.10 $\pm$ 0.03 [a] | 0.09 $\pm$ 0.07 [a] | 0.11 $\pm$ 0.01 [a] |
| Zn | 0.51 $\pm$ 0.05 [a] | 0.56 $\pm$ 0.08 [a] | 0.52 $\pm$ 0.06 [a] | 0.38 $\pm$ 0.02 [a] | 0.52 $\pm$ 0.06 [a] | 0.44 $\pm$ 0.09 [a] | 0.39 $\pm$ 0.01 [a] |
| Fe | 0.34 $\pm$ 0.01 [ab] | 0.34 $\pm$ 0.06 [a] | 0.31 $\pm$ 0.01 [ab] | 0.39 $\pm$ 0.03 [a] | 0.31 $\pm$ 0.01 [ab] | 0.24 $\pm$ 0.03 [b] | 0.32 $\pm$ 0.04 [ab] |
| **Vitamin Content** (mg per 100 g of PW) | | | | | | | |
| Retinol (A) | n.d. | 0.14 $\pm$ 0.01 [a] | n.d. | 0.31 $\pm$ 0.02 [ab] | 0.22 $\pm$ 0.02 [ab] | 0.22 $\pm$ 0.01 [b] | n.d. |
| Thimine (B1) | 0.05 $\pm$ 0.01 [ab] | 0.13 $\pm$ 0.01 [a] | 0.11 $\pm$ 0.01 [ab] | 0.13 $\pm$ 0.01 [ab] | 0.08 $\pm$ 0.01 [ab] | 0.13 $\pm$ 0.02 [b] | 0.06 $\pm$ 0.01 [b] |
| Rboflvin (B2) | 0.08 $\pm$ 0.01 [a] | 0.12 $\pm$ 0.01 [a] | 0.11 $\pm$ 0.01 [a] | 0.13 $\pm$ 0.01 [a] | 0.15 $\pm$ 0.01 [a] | 0.16 $\pm$ 0.01 [a] | 0.11 $\pm$ 0.01 [a] |
| Ascorbic Acid (C) | 43.2 $\pm$ 0.85 [ab] | 25.4 $\pm$ 0.46 [abc] | 32.2 $\pm$ 0.35 [abc] | 42.6 $\pm$ 0.94 [a] | 31.9 $\pm$ 0.83 [bc] | 50.1 $\pm$ 1.98 [c] | 38.2 $\pm$ 0.75 [c] |

Additionally, *Fino de Jete* also showed the lowest fat content. Concerning fibers, they were about 3% of the PW, reaching more than 4% in *Campas*, *Daniela*, *Fino de Jete*, and *Torre1*. Our results, in agreement with other literature data, demonstrated that cherimoya is a tropical fruit with high nutritional value. Indeed, it had low fat content, while containing an amount of sugars and proteins generally higher than other common tropical fruits, including mango [7], kiwi [45], pineapple [46], and papaya [10].

Micronutrients, including minerals and vitamins, are involved in several biochemical processes, and their balanced intake is important to prevent deficiency diseases. Plant foods are important sources of these nutrients [47]. The mineral composition in 100 g of PW is reported in Table 5. K was the most abundant mineral in all the analyzed samples, ranging between 25% and 42% of the total mineral content. Moreover, our analysis showed how Cherimoya, as well as other tropical fruits, is a very reach source of Mg, Ca, and P. In particular, our results showed that the observed seven CVs of Cherimoya had an amount of these micronutrients from two- to four-fold higher than other common edible fruits, such peaches [48,49] and apple [50]. The amount of Na recorded had a mean value of 19.09 $\pm$ 2.02 mg per 100 g of PW; that is higher than that reported for others tropical fruits, such as banana, guava, mango, papaya and pineapple [51]. Concerning micro-minerals, our results suggest that cherimoya fruits are an extraordinary source of Zn, containing two-fold the amount normally present in red currant [51]. Moreover, we recorded high amounts of Mn and Cu, meanwhile the Fe contents were markedly lower. Globally, except for Ca and Zn content, our analysis showed a mineral content comparable to that obtained by Leterme [47]. Finally, concerning the general mineral composition among the different CVs, our results revealed significant differences only in K content, recording the highest content in *Fino de Jete* and lowest one in *Torre2*.

Concerning the quantified vitamins, our analysis revealed that the analyzed fruits are a good source of ascorbic acid, with a mean value equal to 37.66 $\pm$ 8.41 mg per 100 g of PW. On the other hand, we found a great variability among the observed seven CVs. Indeed, *Torre2* was the CV with the highest content of Vitamin C, recording an amount two-fold higher than *Chaffey*.

### 3.4. Nutraceutical Parameters

### 3.4.1. Total Phenolic Content

Polyphenol compounds are the most abundant dietary phytochemicals [40], and several scientific reports demonstrate their positive influence on human health [9]. On the other hand, several biological actions are documented, including antioxidant [52], antinflammatory [53,54], antidiabetic [55], antiproliferative [3,8], antihypertensive [56], and antihyperlipidemic [57] effects. In this work, TPC in the flesh of the seven observed CVs was measured via Folin-Ciocalteu assay (Figure 2). Our analysis revealed that TPC varied between 28.50 $\pm$ 1.92 (*Fino de Jete*) and 174.90 $\pm$ 11.69 (*Chaffey*) mg GAE per 100 g of PW, recording an average value equal to 75.18 $\pm$ 57.94 mg GAE per 100 g PW. The mean value is higher than those reported for the flesh of other tropical fruits with high commercial impact, including kiwi, papaya, mango and avocado [7,10,45,58]. Furthermore, the mean value for TPC in our fruits was 10-fold higher than for cherimoya fruits from Portugal [59], but 3-fold lower than those obtained for the flesh of fruits from Ecuador [60]. The very large range suggests a significant variability among the analyzed genotypes. Although the contribution of reducing compounds different from polyphenols to TPC value cannot be excluded, the minimal differences in the content of protein and sugar (Tables 4 and 5) among the seven observed cherimoya fruits suggests that the observed range in TPC mainly depended on a different content of polyphenols. Among the observed CVs, the highest value was observed for *Chaffey*, followed by *Daniela*. On the other hand, the CVs with the highest commercial impact (*Campas*, *White* and *Fino de Jete*) recorded the lowest TPC. Finally, the TPC of the two local ecotypes, *Torre2* and especially *Torre1*, was considerable, and only lower than that recorded for *Chaffey* and *Daniela*.

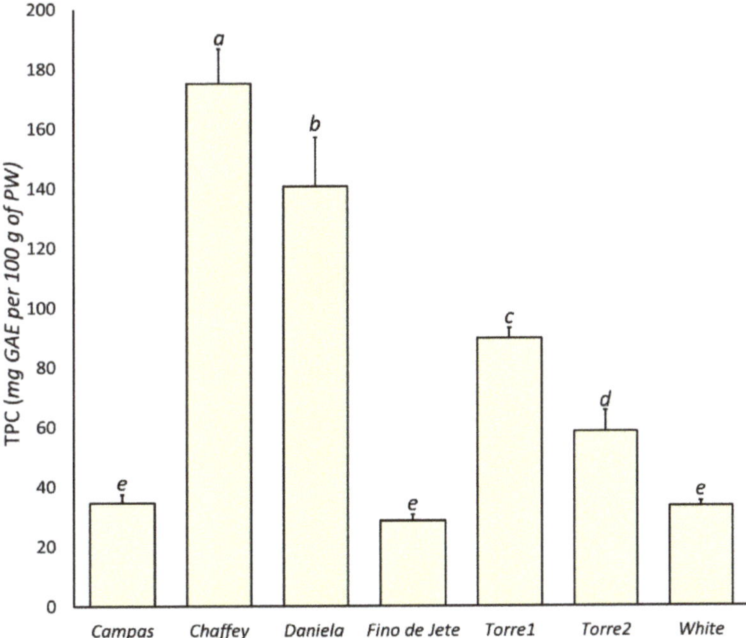

**Figure 2.** The total polyphenolic content (TPC) in the flesh of the nine observed CVs of *Annona cherimola* fruits measured via Folin-Ciocolteau assay. The bars represent mean ± SD. Different lowercase letters on the top of bars indicate significant differences at $p \leq 0.05$ as measured by one-way ANOVA followed by Tukey's multiple range test. The letter "a" denotes the highest value.

In our previous work, we also evaluated the phytochemical composition of leaves obtained from the same observed CVs of *A. cherimola* [3]. Comparing TPC values measured in the leaves with that recorded in this work for the flesh of the respective fruits, we found that for all the observed CVs, the leaves contained more polyphenols than the fruits. Moreover, for the leaves, a different ranking for TPC was observed. In particular, *Torre2*, *White*, and *Fino de Jete* were the CVs with the highest TPC in the leaves; meanwhile, *Daniela* and *Torre1* were those with the lowest [3].

### 3.4.2. Content of Proanthocyanidins

Proanthocyanidins are polyphenols of high molecular weight with documented protective actions for human well-being [61]. In particular, their potential protective effect on the gastrointestinal tract is very interesting. Indeed, thanks both to their high digestive stability at gastrointestinal conditions, and to their reduced intestinal absorption [62–64], PACs may reach high concentrations in the intestinal lumen, producing significant biological effects at the local level. The potential benefit of PACs in chronic intestinal inflammation is supported by numerous studies [9,40,65–70], and epidemiological data show an inverse correlation between food intake rich in proanthocyanidins and the risk of developing colorectal cancer [9,71,72].

We evaluated tPACs in the flesh of the seven observed cherimoya fruits via BL-DMAC assay (Figure 3). We recorded a tPAC content ranging between 10.33 ± 4.51 (*Fino de Jete*) and 51.67 ± 4.04 (*Daniela*) mg PAC-A equivalent per 100 g of PW, with mean values of 28.54 ± 7.98 mg PAC-A equivalent per 100 g of PW. In particular, the highest tPACs was recorded for *Daniela*, followed by *Chaffey*, *Torre2* and *Torre1*; meanwhile, the lowest content was recorded for the *Fino de Jete* and *Campas*.

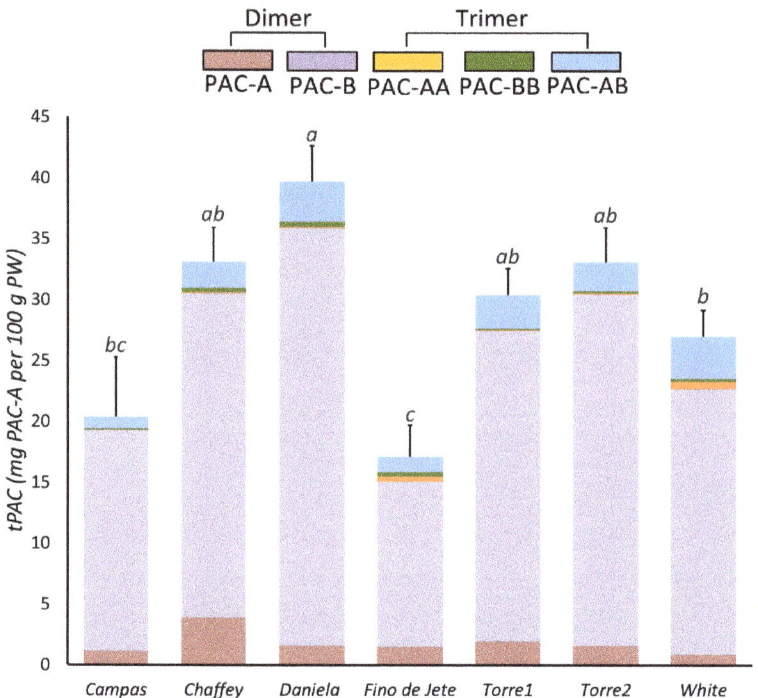



**Figure 3.** Total proanthocyanidin content (tPACs) in the flesh of the nine observed CVs of *Annona cherimola* fruits measured via BL-DMAC assay. The bars represent mean ± SD. Different lowercase letters on the top of the bars indicate significant differences at the $p \leq 0.05$ level as measured by one-way ANOVA followed by Tukey's multiple range test. The letter "a" denotes the highest value. Inside each bar, the different colors indicate the percentage composition of PAC measured by HPLC-MS/MS, as reported in the Materials and Methods section.

Finally, the correlation coefficient ($p = 0.76$), measured by Pearson statistical analysis (Figure 4), suggested that tPCA strongly contributes to the previously measured TPC value. Our analysis is in accordance with García-Salas et al., who evaluated the tPAC and TPC contents in ethanolic extracts from fruits of two different CVs of Cherimoya (*Fino de Jete* and *Campas*) cultivated in Spain [12]. In particular, they showed that cherimoya pulp essentially contains PACs in addition to hydroxytyrosol and traces of luteolin [12].

Concerning the PACs profile, HPLC-MS/MS analysis revealed that in cherimoya fruits, B-type PACs represented about 90% of the total PACs. In particular, *Daniela* displayed the highest percentage, containing more than 94% of B-type PAC; meanwhile, *Chaffey* was the CV with the highest A-type PACs percentage, reaching more than 13%. Even if the greater bioactivity of PAC-A with respect to PAC-B type is well-known [3,9,32,73], the presence of the A-type PAC is very limited in food sources [74,75]. On the other hand, the literature data suggest a higher dosage of PAC-B can exert comparable bioactivity to the PAC-A type [73]. Regarding the polymerization grade of PACs contained in our fruit extracts, we found dimers and a small number of trimers. Our findings are in accordance with Garcia and colleagues, that reported almost exclusively low molecular weight PACs in cherimoya flesh [12]. Although the limited PAC bioavailability, experimental scientific data indicate that their intestinal absorption is inversely related to the polymerization degree [62]. Consequently, dimers and trimers of PAC may also poorly absorbed at the intestinal level [62,76]. Our results would therefore indicate that the cherimoya PAC fraction may be at least partially bioavailable.

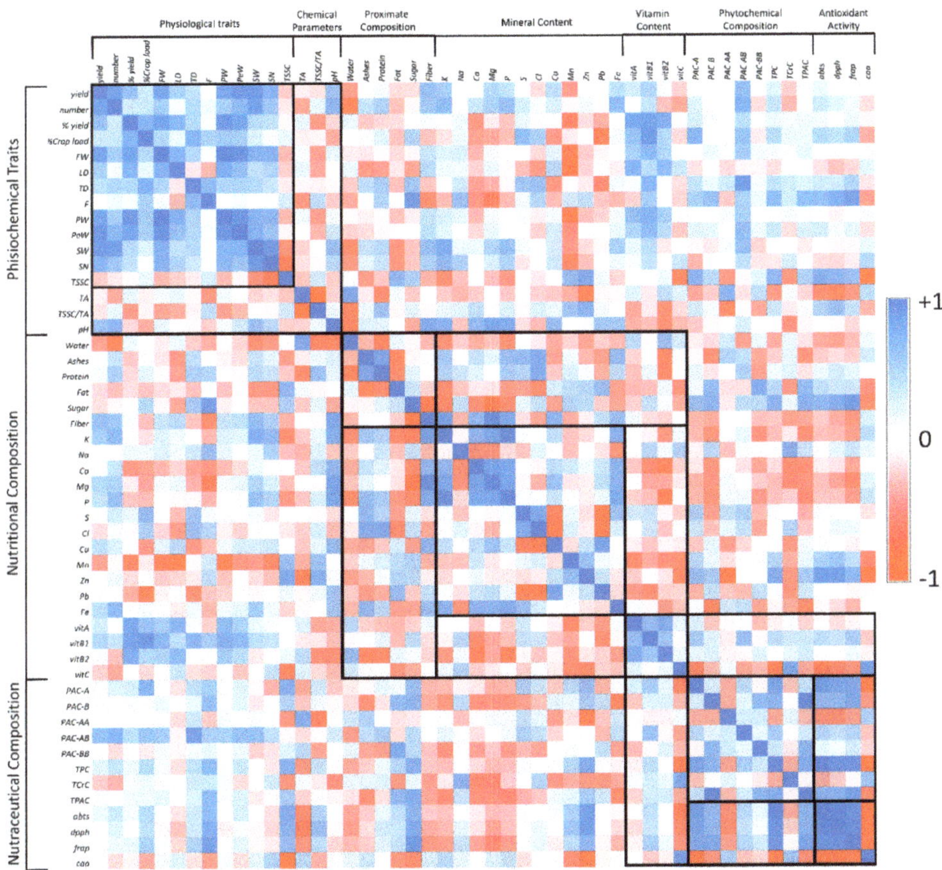

**Figure 4.** A Pearson's correlation heat map displaying the correlation coefficient (ρ) based on physiochemical, nutritional, and nutraceutical data of the seven observed cultivars of *Annona cherimola* fruits. Different colors represent the negative (red) to positive (blue) correlation between two different parameters.

### 3.4.3. Content of Carotenoids

Several scientific studies probe how carotenoid intake contributes to preventing human diseases related to oxidative stress [77]. On the one hand, animals are not able to synthesize carotenoids, and plant foods constitute the major carotenoid sources in the human diet [7,77]. The TCC of the seven observed CVs of *A. cherimola* fruits is reported in Figure 5. Our results showed that cherimoya fruits contain very low amounts of carotenoids. Indeed, the TCC ranged between 10.33 ± 4.51 (*Torre1*) and 51.67 ± 4.04 (*Torre2*) µg β-carotene per 100 g of PW, with an average value of 29.22 ± 12.47 µg β-carotene per 100 g of PW. This value is about 100-fold less than that recorded for mango and papaya fruits [7,10]. On the other hand, our results were in accordance with those listed in the USDA National Nutrient Database [78].

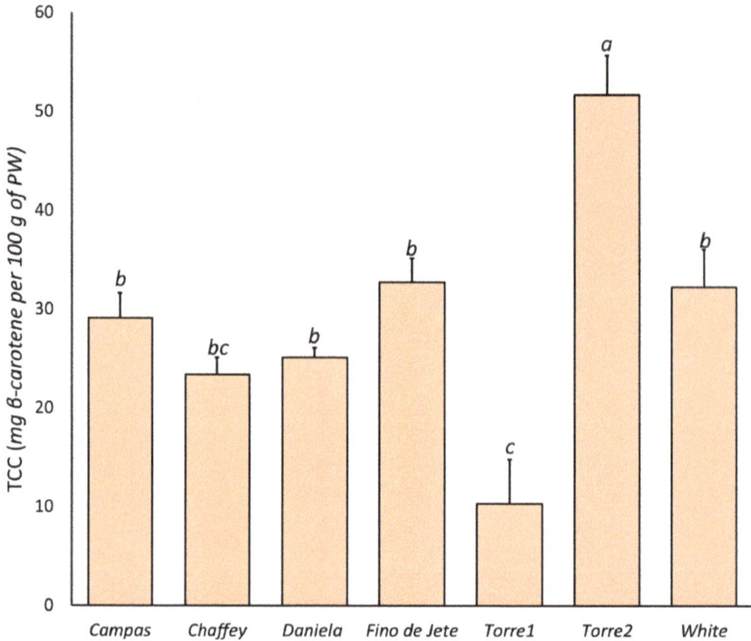

**Figure 5.** The Total Carotenoid Content (TCC) in the flesh of the nine observed CVs of *Annona cherimola* fruits. The bars represent mean±SD. Different lowercase letters on the top of bars indicate significant differences at $p \leq 0.05$ as measured by one-way ANOVA followed by Tukey's multiple range test. The letter "a" denotes the highest value.

### 3.4.4. Antioxidant Properties

Phytocomponents display various biological properties, and the determination of the potential bioactivity of plant extracts may contribute to their valorization for food fortification, but also be of use in the cosmetic and pharmaceutical fields [3,9]. Frequently, the bioactivity of phytochemicals is related to their antioxidative properties, not only preventing oxidative stress but also being useful for the modulation of important redox-dependent cellular functions [48,79].

In this work, the antioxidant properties of the ethanolic extracts of the flesh of the observed cherimoya fruits were evaluated in in solution assays and in a cell-based lipid peroxidation model. Regarding in solution assays, ABTS and DPPH were used to measure the radical scavenging activity; meanwhile, FRAP was used to evaluate the metal-reducing activity (Figure 6, Panel A). The mean values for the radical scavenging activity evaluated by DPPH and ABTS assay were $363.11 \pm 153.59$ and $228.75 \pm 90.52$ mmol TE per 100 g PW, respectively. A lower average value ($1.36 \pm 0.59$ mmol TE per 100 g PW) was measured for the metal-reducing activity via FRAP assay. Peculiar characteristics of the reaction mixtures of the different assays and specific differences in the electronic transfer mechanism may explain the different antioxidant activities recorded [80]. Despite the variability evaluated in the antioxidant activity, the trend among the analyzed CVs was not influenced by the different assays, as suggested by the positive correlation (Figure 4) between the values obtained from the three assays ($\rho_{ABTS/DPPH} = 0.968$; $]\rho_{ABTS/FRAP} = 0.917$; $\rho_{DPPH/FRAP} = 0.949$). On the other hand, the obtained results highlight a significant variability in the nutraceutical potential of the analyzed CVs. *Chaffey* and *Daniela* always showed the highest antioxidant activities; meanwhile, *White*, *Fino de Jete* and *Campas* displayed the lowest ones, both in terms of radical scavenging and reducing activity.

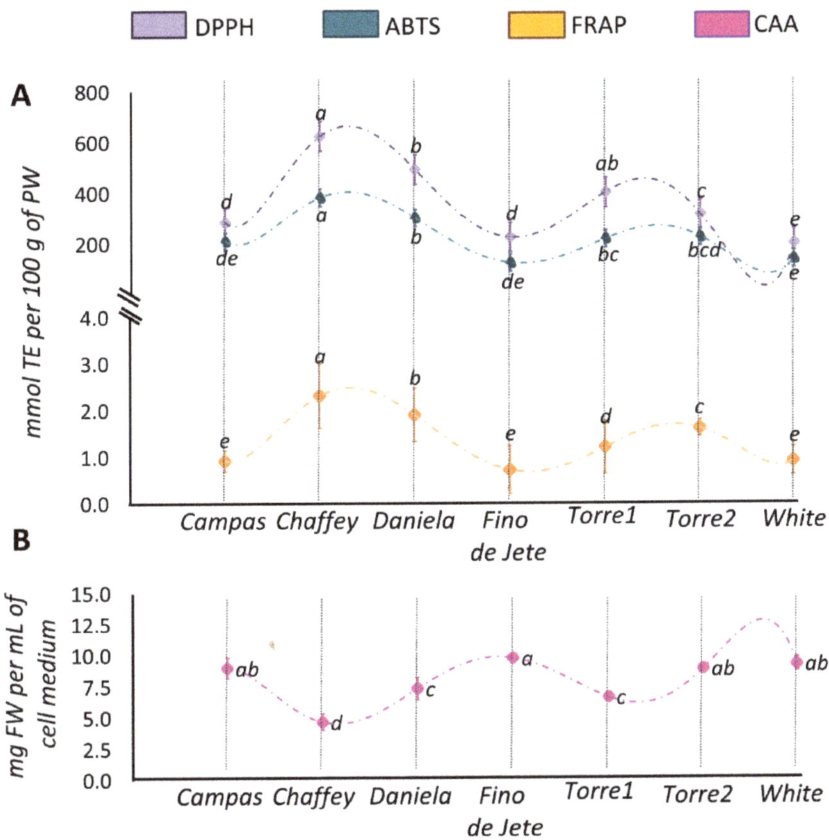



**Figure 6.** The antioxidant activities of ethanolic extracts of the flesh of the seven observed CVs of *Annona cherimola*. Panel (**A**) shows values measured by the radical scavenging (DPPH = violet; ABTS = light blue) and metal-reducing (FRAP = yellow) assays. Panel (**B**) shows THE CAA$_{50}$ value. Within the same series, different lowercase letters indicate significant difference at the $p \leq 0.05$ level as measured by one-way ANOVA followed by Tukey's multiple range test. The letter "a" denotes the value.

Additionally, the obtained values were positively correlated with TPC ($\rho_{TCP/DPPH}$ = 0.983; $\rho_{TCP/ABTS}$ = 0.940; $\rho_{TCP/FRAP}$ = 0.928) (Figure 4), indicating that polyphenols contribute almost exclusively to the redox-active properties. Moreover, tPACs was also strongly correlated with FRAP ($\rho_{TPAC/FRAP}$ = 0.844), but it found a lower correlation with ABTS and DPPH ($\rho_{TPAC/ABTS}$ = 0.654; $\rho_{TPAC/DPPH}$ = 0.664). The higher correlation found between FRAP and tPAC may be explained by the peculiar structural characteristics of proanthocyanidins, making it easier to bind metal ions thanks the presence of free meta-oriented hydroxyl groups [3,7].

Although in solution assays are widely employed to preliminarily evaluate the antioxidant capacities, they cannot measure the antioxidant activity in a biological environment [40,80]. For this purpose, cellular-based lipid peroxidation models evaluate the potential ability of redox-active compounds to interact with biological membranes [39,81,82]. Therefore, they represent interesting alternatives to in vivo models, which instead may be expensive, unethical, and not easy to use [39]. In particular, the CAA assay is a very biologically relevant method because it also takes into account the uptake, metabolism, and location of antioxidant compounds within cells [39,81]. The antioxidant activity of the ethanolic extracts of the fruits from the seven observed CVs of *A. cherimola* expressed as $CAA_{50}$ is displayed in (Figure 6, Panel B). The average $CAA_{50}$ value was 7.94 mg $\pm$ 1.81 of the PW per mL of cell medium was recorded. The obtained values are in the same range as those determined by Wolfe et al. under the same experimental conditions for hydrophilic extracts of other fresh fruits [81]. Among the observed CVs, a little variability in term of $CAA_{50}$ was recorded. In particular, *Chaffey* displayed the highest activity, followed by *Daniela* and *Torre1*. Although in-solution assays are not always predictive of the antioxidant capacity in biological models, in our experimental conditions, the antioxidant activity of CAA is positively correlated with the redox-active properties evaluated by in solution assays ($\rho_{CAA/ABTS}$ = $-0.943$; $\rho_{CAA/DPPH}$ = $-0.879$; $\rho_{CAA/FRAP}$ = $-0.818$) (Figure 4). Furthermore, the strong correlation between the $CAA_{50}$ values and TPC ($\rho_{CAA/TPC}$ = $-0.923$) and the lower correlation with tPACs ($\rho_{CAA/TPAC}$ = $-0.608$) of the tested extracts suggested that the main contribution to CAA was not mainly given by the PACs, but by other polyphenol compounds. This result is not surprising, considering the low capacity of polyphenolic polymers to cross the cellular membrane [62].

*3.5. Cultivar Discrimination via Principal Component Analysis and HeatMap Cluster Analysis*

The Principal Component Analysis (PCA), calculated on the data matrices related to pomological, physiochemical, nutritional and nutraceutical values previously measured, allowed for the discrimination between the different fruits of the seven observed CVs of *A. cherimola* (Figure 7). In particular, PCA explained 27.32% and 51.73% of the total variance, respectively for PC1 and PC2. Positive factor scores discriminated *Fino de Jete* and *Daniela* from others CVs. In particular, *Fino de Jete* is the best CV for the highest values related to the most important commercial parameters, such as FW, LD, TD, F, and PW, but it also showed the best nutritional profile among the different CVs, having a good vitamin and mineral content. On the other hand, *Daniela* showed the best phytochemical profile and the highest antioxidant properties while having intermediate pomological, physicochemical, and nutritional values. *Campas* was completely separated from other CVs for positive PC2 and negative PC1. In particular, the separation is mainly due to the high mineral content, and to poor nutraceutical properties and very poor pomological traits. Finally, *Torre1*, *Torre2*, *White* and *Chaffey* were grouped for negative PC2. Especially, *Torre1* and *White* had also negative PC1 due to their similar pomological characteristics. On the other hand, positive PC1 and negative PC2 factor scores grouped *Torre2* and *Chaffey* for their particular PAC composition. However, *Chaffey* also had nutritional and nutraceutical properties better than *Torre2*.

The HeatMap coupled with Hierarchical Clustering Analysis confirmed the separation performed by PCA (Figure 8). In particular, *Fino de Jete* was completely separated from other CVs due to high values recorded both for some of the pomological traits, such as FW, LD, TD, PW, PeW and SN, and for the highest vitamin content. On the other hand, *White* were really far from *Fino de Jete* because the lowest values recorded for all the pomological traits of fruits, and also for the low vitamin and phytochemical content. For the same reason, *Campas* was very close to *White*. Clustering analysis revealed that the two local ecotypes, *Torre1* and *Torre2*, had not only similar pomological parameters, but they also displayed a comparable antioxidant activity both in solution and in cellular models. This proximity may also be explained by their similar PAC profile. Finally, *Daniela* and *Chaffey* take place in an intermediate position within the clustering due to their acceptable values recorded both for pomological traits, nutritional values, and nutraceutical properties.

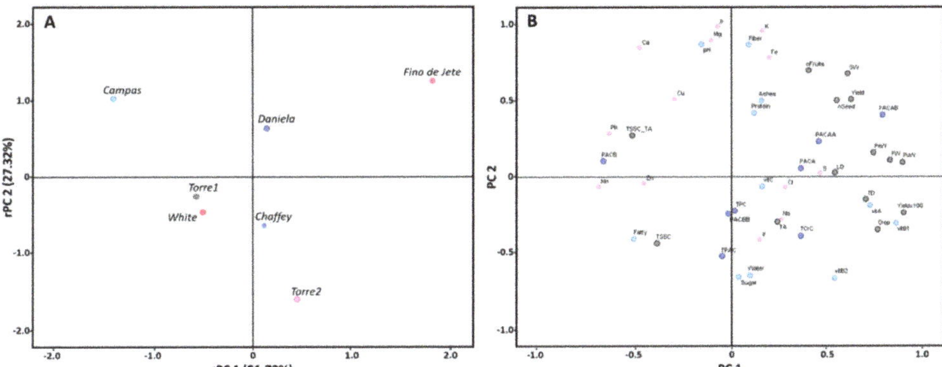

**Figure 7.** Scatter plot of the principal components factor scores of the seven observed CVs of *Annona cherimola* fruits. Panel (**A**) shows the clear separation among the different CVs; meanwhile, Panel (**B**) reports the chemical portioning of the compounds.

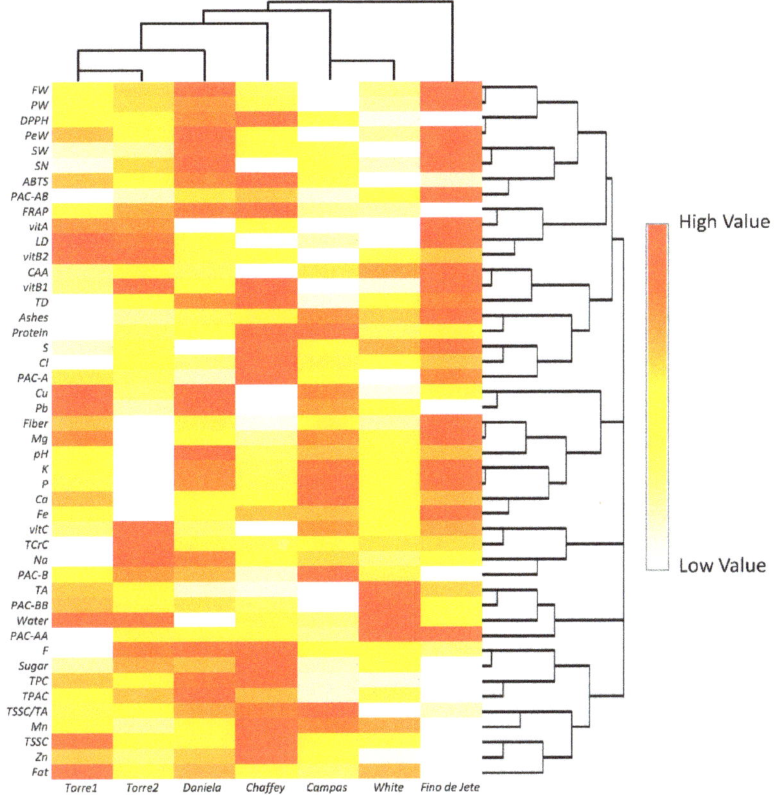

**Figure 8.** Hierarchical clustering analysis and heatmap visualization of the physiochemical, nutritional, and nutraceutical data of the seven observed cultivars of *Annona cherimola* fruits. For each row, diverse colors indicate differences between the values measured for each parameter among the seven cultivars.

## 4. Conclusions

In the present work, we demonstrated the overall high quality of cherimoya fruits harvested from plants grown in Sicily, in terms of pomological, physiochemical, nutritional, and nutraceutical attributes. Our results showed great variability among the seven observed CVs and may contribute to better define the potential commercial impact of the different CVs. In particular, our analysis showed *Fino de Jete* as being a commercially appreciated CV for its pomological and physiochemical attributes; it also had high nutritional values. On the other hand, the local CV *Daniela*, together to good commercial attributes, also displayed good nutraceutical properties. Concerning *White*, a CV less requested on the market, in addition to having low pomological attributes, it had reduced nutritional and nutraceutical values. Finally, the two local ecotypes, *Torre1* and *Torre2*, had qualitative attributes comparable to those measured for the international CVs.

**Author Contributions:** Conceptualization, C.G.; Data curation, C.G., G.M., G.S., and V.F.; Formal analysis, G.M., E.P., G.G., A.P., G.S., and V.F.; Funding acquisition, C.G.; Investigation, G.M. and E.P.; Methodology, G.M.; Project administration, C.G.; Resources, C.G.; Software, G.M. and G.G.; Supervision, C.G., G.M. and V.F.; Validation, C.G., G.M. and E.P.; Visualization, G.M. and G.G.; Writing—original draft, C.G. and G.M.; Writing—review & editing, C.G. and G.M. All authors have read and agreed to the published version of the manuscript.

**Funding:** This research received no external funding.

**Acknowledgments:** The authors would like to thank Vivai Torre s.r.l. for providing samples and for their kind hospitality. Moreover, we would like to express our gratitude to Valentina Sciacca and Stefano Puccio for their technical support.

**Conflicts of Interest:** The authors declare no conflict of interest.

## References

1. Quílez, A.; Fernández-Arche, M.; García-Giménez, M.; De La Puerta, R. Potential therapeutic applications of the genus Annona: Local and traditional uses and pharmacology. *J. Ethnopharmacol.* **2018**, *225*, 244–270. [CrossRef] [PubMed]
2. Viera-Arroyo, W.; Campaña, D.; Castro, S.; Vásquez, W.; Viteri, P.; Zambrano, J.L. Effectiveness of the arbuscular mycorrhizal fungi use in the cherimoya (Annona cherimola Mill.) seedlings growth. *Acta Agronómica* **2017**, *66*, 993–1002. [CrossRef]
3. Mannino, G.; Gentile, C.; Porcu, A.; Agliassa, C.; CaraDonna, F.; Bertea, C.M. Chemical Profile and Biological Activity of Cherimoya (Annona cherimola Mill.) and Atemoya (Annona atemoya) Leaves. *Molecules* **2020**, *25*, 2612. [CrossRef] [PubMed]
4. FAO. *The State of Food and Agriculture, 1996*; Food and Agriculture Organization of United Nations: Rome, Italy, 1993; pp. 85–90.
5. Pino, J.A.; Roncal, E. Characterisation of odour-active compounds in cherimoya (Annona cherimolaMill.) fruit. *Flavour Fragr. J.* **2016**, *31*, 143–148. [CrossRef]
6. Pareek, S.; Yahia, E.M.; Pareek, O.; Kaushik, R. Postharvest physiology and technology of Annona fruits. *Food Res. Int.* **2011**, *44*, 1741–1751. [CrossRef]
7. Gentile, C.; Di Gregorio, E.; Di Stefano, V.; Mannino, G.; Perrone, A.; Avellone, G.; Sortino, G.; Inglese, P.; Farina, V. Food quality and nutraceutical value of nine cultivars of mango (*Mangifera indica* L.) fruits grown in Mediterranean subtropical environment. *Food Chem.* **2019**, *277*, 471–479. [CrossRef]
8. Gentile, C.; Reig, C.; Corona, O.; Todaro, A.; Mazzaglia, A.; Perrone, A.; Gianguzzi, G.; Agustí, M.; Farina, V. Pomological Traits, Sensory Profile and Nutraceutical Properties of Nine Cultivars of Loquat (Eriobotrya japonica Lindl.) Fruits Grown in Mediterranean Area. *Plant Foods Hum. Nutr.* **2016**, *71*, 330–338. [CrossRef]
9. Mannino, G.; Di Stefano, V.; Lauria, A.; Pitonzo, R.; Gentile, C. Vaccinium macrocarpon (Cranberry)-Based Dietary Supplements: Variation in Mass Uniformity, Proanthocyanidin Dosage and Anthocyanin Profile Demonstrates Quality Control Standard Needed. *Nutrients* **2020**, *12*, 992. [CrossRef]
10. Farina, V.; Tinebra, I.; Perrone, A.; Sortino, G.; Palazzolo, E.; Mannino, G.; Gentile, C. Physicochemical, Nutraceutical and Sensory Traits of Six Papaya (*Carica papaya* L.) Cultivars Grown in Greenhouse Conditions in the Mediterranean Climate. *Agronomy* **2020**, *10*, 501. [CrossRef]
11. Compositional Evaluation of Annona cherimoya (Custard Apple) Fruit. *Trends Appl. Sci. Res.* **2008**, *3*, 216–220. [CrossRef]
12. García-Salas, P.; Gómez-Caravaca, A.M.; Morales-Soto, A.; Segura-Carretero, A.; Fernández-Gutiérrez, A. Identification and quantification of phenolic and other polar compounds in the edible part of Annona cherimola and its by-products by HPLC-DAD-ESI-QTOF-MS. *Food Res. Int.* **2015**, *78*, 246–257. [CrossRef] [PubMed]
13. Santos, S.A.O.; Vilela, C.; Camacho, J.F.; Cordeiro, N.; Gouveia, M.M.C.; Freire, C.S.R.; Silvestre, A.J.D. Profiling of lipophilic and phenolic phytochemicals of four cultivars from cherimoya (Annona cherimola Mill.). *Food Chem.* **2016**, *211*, 845–852. [CrossRef] [PubMed]

14. Rabêlo, S.V.; Costa, E.V.; Barison, A.; Dutra, L.M.; Nunes, X.P.; Tomaz, J.C.; Oliveira, G.G.; Lopes, N.P.; Santos, M.D.F.C.; Almeida, J.R.G.D.S. Alkaloids isolated from the leaves of atemoya (Annona cherimola×Annona squamosa). *Rev. Bras. Farm.* **2015**, *25*, 419–421. [CrossRef]

15. Chen, C.-Y.; Chang, F.-R.; Pan, W.-B.; Wu, Y.-C. Four alkaloids from Annona cherimola. *Phytochemistry* **2001**, *56*, 753–757. [CrossRef]

16. Haykal, T.; Nasr, P.; Hodroj, M.H.; Taleb, R.I.; Sarkis, R.; Moujabber, M.N.E.; Rizk, S. Annona cherimola Seed Extract Activates Extrinsic and Intrinsic Apoptotic Pathways in Leukemic Cells. *Toxins* **2019**, *11*, 506. [CrossRef]

17. Ammoury, C.; Younes, M.; El Khoury, M.; Hodroj, M.H.; Haykal, T.; Nasr, P.; Sily, M.; Taleb, R.I.; Sarkis, R.; Khalife, R.; et al. The pro-apoptotic effect of a Terpene-rich Annona cherimola leaf extract on leukemic cell lines. *BMC Complement. Altern. Med.* **2019**, *19*, 1–10. [CrossRef]

18. Wele, A.; Zhang, Y.; Brouard, J.-P.; Pousset, J.-L.; Bodo, B. Two cyclopeptides from the seeds of Annona cherimola. *Phytochemistry* **2005**, *66*, 2376–2380. [CrossRef] [PubMed]

19. Manríquez, D.A.; Muñoz-Robredo, P.; Gudenschwager, O.; Robledo, P.; Defilippi, B.G. Development of flavor-related metabolites in cherimoya (Annona cherimola Mill.) fruit and their relationship with ripening physiology. *Postharvest Biol. Technol.* **2014**, *94*, 58–65. [CrossRef]

20. Galarce-Bustos, O.; Fernández-Ponce, M.T.; Montes, A.; Pereyra, C.; Casas, L.; Mantell, C.; Aranda, M.; Cardoso, L.C.; Serrano, C.M. Usage of supercritical fluid techniques to obtain bioactive alkaloid-rich extracts from cherimoya peel and leaves: Extract profiles and their correlation with antioxidant properties and acetylcholinesterase and $\alpha$-glucosidase inhibitory activities. *Food Funct.* **2020**, *11*, 4224–4235. [CrossRef]

21. Macuer-Guzmán, J.; Bernal, G.; Jamett-Díaz, F.; Ramírez-Rivera, S.; Ibáñez, C. Selective and Apoptotic Action of Ethanol Extract of Annona cherimola Seeds against Human Stomach Gastric Adenocarcinoma Cell Line AGS. *Plant Foods Hum. Nutr.* **2019**, *74*, 322–327. [CrossRef]

22. Calzada, F.; Correa-Basurto, J.; Barbosa, E.; Mendez-Luna, D.; Yépez-Mulia, L. Antiprotozoal Constituents from Annona cherimola Miller, a Plant Used in Mexican Traditional Medicine for the Treatment of Diarrhea and Dysentery. *Pharmacogn. Mag.* **2017**, *13*, 148–152. [PubMed]

23. Vasarri, M.; Barletta, E.; Vinci, S.; Ramazzotti, M.; Francesconi, A.; Manetti, F.; Degl'Innocenti, D. *Annona cherimola* Miller Fruit as a Promising Candidate against Diabetic Complications: An In Vitro Study and Preliminary Clinical Results. *Foods* **2020**, *9*, 1350. [CrossRef] [PubMed]

24. Yemm, E.W.; Willis, A.J. The estimation of carbohydrates in plant extracts by anthrone. *Biochem. J.* **1954**, *57*, 508–514. [CrossRef][PubMed]

25. Jung, S.; Rickert, D.A.; Deak, N.A.; Aldin, E.D.; Recknor, J.; Johnson, L.A.; Murphy, P.A. Comparison of kjeldahl and dumas methods for determining protein contents of soybean products. *J. Am. Oil Chem. Soc.* **2003**, *80*, 1169–1173. [CrossRef]

26. Williams, S. *Official Methods of Analysis*; Association of Official Agricultural Chemists: Rockville, MD, USA, 1984; p. 14.

27. Vigliante, I.; Mannino, G.; Maffei, M.E. Chemical Characterization and DNA Fingerprinting of Griffonia simplicifolia Baill. *Molecules* **2019**, *24*, 1032. [CrossRef] [PubMed]

28. Farina, V.; Gianguzzi, G.; D'Asaro, A.; Mazzaglia, A.; Palazzolo, E. Fruit production and quality evaluation of four litchi cultivars (Litchi chinensis Sonn.) grown in Mediterranean climate. *Fruits* **2017**, *72*, 203–211. [CrossRef]

29. Zhao, P.J.; Zhang, B.C.; WANG, L.; LI, G. Determination of phosphorus content in foods by vanadium molybdate yellow colorimetric method. *Shandong Agric. Sci.* **2009**, *2*.

30. Ollilainen, V.; Vahteristo, L.; Uusi-Rauva, A.; Varo, P.; Koivistoinen, P.; Huttunen, J. The HPLC Determination of Total Thiamin (Vitamin B1) in Foods. *J. Food Compos. Anal.* **1993**, *6*, 152–165. [CrossRef]

31. Bueno-Solano, C.; López-Cervantes, J.; Campas-Baypoli, O.N.; Cortez-Rocha, M.O.; Casillas-Hernández, R.; Milan-Carrillo, J.; Sánchez-Machado, D.I. Quantitative HPLC Analysis of Riboflavin and Aromatic Amino Acids in Three Forms of Shrimp Hydrolysates. *J. Liq. Chromatogr. Relat. Technol.* **2009**, *32*, 3009–3024. [CrossRef]

32. Mannino, G.; Gentile, C.; Maffei, M.E. Chemical partitioning and DNA fingerprinting of some pistachio (*Pistacia vera* L.) varieties of different geographical origin. *Phytochemistry* **2019**, *160*, 40–47. [CrossRef]

33. Mannino, G.; Nerva, L.; Gritli, T.; Novero, M.; Fiorilli, V.; Bacem, M.; Bertea, C.; Lumini, E.; Chitarra, W.; Balestrini, R.M. Effects of Different Microbial Inocula on Tomato Tolerance to Water Deficit. *Agronomy* **2020**, *10*, 170. [CrossRef]

34. Prior, R.L.; Fan, E.; Ji, H.; Howell, A.; Nio, C.; Payne, M.J.; Reed, J. Multi-laboratory validation of a standard method for quantifying proanthocyanidins in cranberry powders. *J. Sci. Food Agric.* **2010**, *90*, 1473–1478. [CrossRef] [PubMed]

35. Vigliante, I.; Mannino, G.; Maffei, M.E. OxiCyan®, a phytocomplex of bilberry (Vaccinium myrtillus) and spirulina (Spirulina platensis), exerts both direct antioxidant activity and modulation of ARE/Nrf2 pathway in HepG2 cells. *J. Funct. Foods* **2019**, *61*, 103508. [CrossRef]

36. Re, R.; Pellegrini, N.; Proteggente, A.; Pannala, A.; Yang, M.; Rice-Evans, C. Antioxidant activity applying an improved ABTS radical cation decolorization assay. *Free Radic. Biol. Med.* **1999**, *26*, 1231–1237. [CrossRef]

37. Mishra, K.; Ojha, H.; Chaudhury, N.K. Estimation of antiradical properties of antioxidants using DPPH assay: A critical review and results. *Food Chem.* **2012**, *130*, 1036–1043. [CrossRef]

38. Benzie, I.F.F.; Strain, J.J. The ferric reducing ability of plasma (FRAP) as a measure of "antioxidant power": The FRAP assay. *Anal. Biochem.* **1996**, *239*, 70–76. [CrossRef]

39. Wolfe, K.L.; Liu, R.H. Cellular Antioxidant Activity (CAA) Assay for Assessing Antioxidants, Foods, and Dietary Supplements. *J. Agric. Food Chem.* **2007**, *55*, 8896–8907. [CrossRef]
40. Mannino, G.; Perrone, A.; Campobenedetto, C.; Schittone, A.; Bertea, C.M.; Gentile, C. Phytochemical profile and antioxidative properties of Plinia trunciflora fruits: A new source of nutraceuticals. *Food Chem.* **2020**, *307*, 125515. [CrossRef]
41. Silva, E.P.; Boas, E.V.D.B.V.; Xisto, A.L.P.R. Characterization and development of marolo (Annona crassiflora, Mart.). *Food Sci. Technol.* **2013**, *33*, 666–675. [CrossRef]
42. Andrés-Agustín, J.; González-Andrés, F.; Nieto-Angel, R.; Barrientos-Priego, A. Morphometry of the organs of cherimoya (Annona cherimola Mill.) and analysis of fruit parameters for the characterization of cultivars, and Mexican germplasm selections. *Sci. Hortic.* **2006**, *107*, 337–346. [CrossRef]
43. Crisosto, C.; Crisosto, G.; Bowerman, E. Understanding consumer acceptance of peach, nectarine and plum cultivars. *Acta Hortic.* **2003**, 115–119. [CrossRef]
44. Morton, J.F. *Fruits of Warm Climates*; J.F. Morton: Miami, FL, USA, 1987; ISBN 0961018410.
45. Passafiume, R.; Gaglio, R.; Sortino, G.; Farina, V. Effect of Three Different *Aloe vera* Gel-Based Edible Coatings on the Quality of Fresh-Cut "Hayward" Kiwifruits. *Foods* **2020**, *9*, 939. [CrossRef] [PubMed]
46. Hossain, F.; Akhtar, S.; Anwar, M. Nutritional Value and Medicinal Benefits of Pineapple. *Int. J. Nutr. Food Sci.* **2015**, *4*, 84. [CrossRef]
47. Leterme, P.; Buldgen, A.; Estrada, F.; Londoño, A.M. Mineral content of tropical fruits and unconventional foods of the Andes and the rain forest of Colombia. *Food Chem.* **2006**, *95*, 644–652. [CrossRef]
48. Gil Muñoz, M.I.; Tomás-Barberán, F.A.; Hess-Pierce, B.; Kader, A.A. Antioxidant Capacities, Phenolic Compounds, Carotenoids, and Vitamin C Contents of Nectarine, Peach, and Plum Cultivars from California. *J. Agric. Food Chem.* **2002**, *50*, 4976–4982. [CrossRef]
49. Durst, R.W.; Weaver, G.W. Nutritional content of fresh and canned peaches. *J. Sci. Food Agric.* **2013**, *93*, 593–603. [CrossRef]
50. Podsędek, A.; Wilska-Jeszka, J.; Anders, B.; Markowski, J. Compositional characterisation of some apple varieties. *Eur. Food Res. Technol.* **2000**, *210*, 268–272. [CrossRef]
51. Sanchezmoreno, C.; De Pascual-Teresa, S.; De Ancos, B.; Cano, M.P. Nutritional Values of Fruits. *Handb. Fruits Fruit Process.* **2007**, 29–43. [CrossRef]
52. Zhang, H.; Tsao, R. Dietary polyphenols, oxidative stress and antioxidant and anti-inflammatory effects. *Curr. Opin. Food Sci.* **2016**, *8*, 33–42. [CrossRef]
53. Shen, Y.; Zhang, H.; Cheng, L.; Wang, L.; Qian, H.; Qi, X. In vitro and in vivo antioxidant activity of polyphenols extracted from black highland barley. *Food Chem.* **2016**, *194*, 1003–1012. [CrossRef]
54. Mannino, G.; CaraDonna, F.; Cruciata, I.; Lauria, A.; Perrone, A.; Gentile, C. Melatonin reduces inflammatory response in human intestinal epithelial cells stimulated by interleukin-1β. *J. Pineal Res.* **2019**, *67*, e12598. [CrossRef] [PubMed]
55. Umeno, A.; Horie, M.; Murotomi, K.; Nakajima, Y.; Yoshida, Y. Antioxidative and Antidiabetic Effects of Natural Polyphenols and Isoflavones. *Molecules* **2016**, *21*, 708. [CrossRef] [PubMed]
56. Rodrigo, R.; Gil, D.; Miranda-Merchak, A.; Kalantzidis, G. Antihypertensive role of polyphenols. *Advances in Clinical Chemistry* **2012**, *58*, 225–254. [CrossRef]
57. Shetgiri, P.P.; Darji, K.K.; D'mello, P.M. Evaluation of antioxidant and antihyperlipidemic activity of extracts rich in polyphenols. *Int. J. Phytomedicine* **2010**, *2*, 3.
58. Campos, D.; Teran-Hilares, F.; Chirinos, R.; Aguilar-Galvez, A.; García-Ríos, D.; Pacheco-Avalos, A.; Pedreschi, R. Bioactive compounds and antioxidant activity from harvest to edible ripeness of avocado cv. Hass (Persea americana) throughout the harvest seasons. *Int. J. Food Sci. Technol.* **2020**, *55*, 2208–2218. [CrossRef]
59. Albuquerque, T.G.; Santos, F.; Sanches-Silva, A.; Oliveira, M.B.; Bento, A.C.; Costa, H. Nutritional and phytochemical composition of Annona cherimola Mill. fruits and by-products: Potential health benefits. *Food Chem.* **2016**, *193*, 187–195. [CrossRef]
60. Vasco, C.; Ruales, J.; Kamal-Eldin, A. Total phenolic compounds and antioxidant capacities of major fruits from Ecuador. *Food Chem.* **2008**, *111*, 816–823. [CrossRef]
61. Dixon, R.A.; Xie, D.-Y.; Sharma, S.B. Proanthocyanidins—A final frontier in flavonoid research? *New Phytol.* **2004**, *165*, 9–28. [CrossRef]
62. Deprez, S.; Mila, I.; Huneau, J.-F.; Tomé, D.; Scalbert, A. Transport of Proanthocyanidin Dimer, Trimer, and Polymer Across Monolayers of Human Intestinal Epithelial Caco-2 Cells. *Antioxid. Redox Signal.* **2001**, *3*, 957–967. [CrossRef]
63. Spencer, J.P.; Chaudry, F.; Pannala, A.S.; Srai, S.; Debnam, E.; A Rice-Evans, C. Decomposition of Cocoa Procyanidins in the Gastric Milieu. *Biochem. Biophys. Res. Commun.* **2000**, *272*, 236–241. [CrossRef]
64. Rios, L.Y.; Bennett, R.N.; A Lazarus, S.; Rémésy, C.; Scalbert, A.; Williamson, G. Cocoa procyanidins are stable during gastric transit in humans. *Am. J. Clin. Nutr.* **2002**, *76*, 1106–1110. [CrossRef] [PubMed]
65. Erlejman, A.G.; Fraga, C.G.; Oteiza, P.I. Procyanidins protect Caco-2 cells from bile acid- and oxidant-induced damage. *Free. Radic. Biol. Med.* **2006**, *41*, 1247–1256. [CrossRef]
66. Erlejman, A.G.; Jaggers, G.; Fraga, C.G.; Oteiza, P.I. TNFα-induced NF-κB activation and cell oxidant production are modulated by hexameric procyanidins in Caco-2 cells. *Arch. Biochem. Biophys.* **2008**, *476*, 186–195. [CrossRef] [PubMed]
67. Gentile, C.; Perrone, A.; Attanzio, A.; Tesoriere, L.; Livrea, M.A. Sicilian pistachio (*Pistacia vera* L.) nut inhibits expression and release of inflammatory mediators and reverts the increase of paracellular permeability in IL-1β-exposed human intestinal epithelial cells. *Eur. J. Nutr.* **2014**, *54*, 811–821. [CrossRef] [PubMed]
68. Amy, H. Cranberry Proanthocyanidins and the Maintenance of Urinary Tract Health. *Crit. Rev. Food Sci. Nutr.* **2002**, *42*, 273–278. [CrossRef]

69. Li, X.-L.; Cai, Y.-Q.; Qin, H.; Wu, Y.-J. Therapeutic effect and mechanism of proanthocyanidins from grape seeds in rats with TNBS-induced ulcerative colitis. *Can. J. Physiol. Pharmacol.* **2008**, *86*, 841–849. [CrossRef]
70. Wang, Y.-H.; Yang, X.-L.; Wang, L.; Cui, M.-X.; Cai, Y.-Q.; Li, X.-L.; Wu, Y.-J. Effects of proanthocyanidins from grape seed on treatment of recurrent ulcerative colitis in rats. *Can. J. Physiol. Pharmacol.* **2010**, *88*, 888–898. [CrossRef]
71. Rossi, M.; Negri, E.; Parpinel, M.; Lagiou, P.; Bosetti, C.; Talamini, R.; Montella, M.; Giacosa, A.; Franceschi, S.; La Vecchia, C. Proanthocyanidins and the risk of colorectal cancer in Italy. *Cancer Causes Control* **2009**, *21*, 243–250. [CrossRef]
72. Prior, R.L.; Gu, L. Occurrence and biological significance of proanthocyanidins in the American diet. *Phytochemistry* **2005**, *66*, 2264–2280. [CrossRef]
73. Beecher, G.R. Proanthocyanidins: Biological activities associated with human health. *Pharm. Biol.* **2004**, *42*, 2–20. [CrossRef]
74. Gu, L.; Kelm, M.A.; Hammerstone, J.F.; Beecher, G.; Holden, J.; Haytowitz, D.; Gebhardt, S.; Prior, R.L. Concentrations of proanthocyanidins in common foods and estimations of normal consumption. *J. Nutr.* **2004**, *134*, 613–617. [CrossRef] [PubMed]
75. Santos-Buelga, C.; Scalbert, A. Proanthocyanidins and tannin-like compounds–nature, occurrence, dietary intake and effects on nutrition and health. *J. Sci. Food Agric.* **2000**, *80*, 1094–1117. [CrossRef]
76. Gentile, C.; Allegra, M.; Angileri, F.; Pintaudi, A.M.; Livrea, M.A.; Tesoriere, L. Polymeric proanthocyanidins from Sicilian pistachio (*Pistacia vera* L.) nut extract inhibit lipopolysaccharide-induced inflammatory response in RAW 264.7 cells. *Eur. J. Nutr.* **2012**, *51*, 353–363. [CrossRef] [PubMed]
77. Rao, A.V.; Rao, L. Carotenoids and human health. *Pharmacol. Res.* **2007**, *55*, 207–216. [CrossRef] [PubMed]
78. Review of USDA National Nutrient Database for Standard Reference, Release 24 and Dietary Supplement Ingredient Database, Release 2. *J. Agric. Food Inf.* **2012**, *13*, 358–359. [CrossRef]
79. Virgili, F.; Marino, M. Regulation of cellular signals from nutritional molecules: A specific role for phytochemicals, beyond antioxidant activity. *Free Radic. Biol. Med.* **2008**, *45*, 1205–1216. [CrossRef]
80. Thaipong, K.; Boonprakob, U.; Crosby, K.; Cisneros-Zevallos, L.; Byrne, D.H. Comparison of ABTS, DPPH, FRAP, and ORAC assays for estimating antioxidant activity from guava fruit extracts. *J. Food Compos. Anal.* **2006**, *19*, 669–675. [CrossRef]
81. Wolfe, K.L.; Kang, X.; He, X.; Dong, M.; Zhang, Q.; Liu, R.H. Cellular Antioxidant Activity of Common Fruits. *J. Agric. Food Chem.* **2008**, *56*, 8418–8426. [CrossRef]
82. Allegra, M.; Gentile, C.; Tesoriere, L.; Livrea, M.A. Protective effect of melatonin against cytotoxic actions of malondialdehyde: An in vitro study on human erythrocytes. *J. Pineal Res.* **2002**, *32*, 187–193. [CrossRef]

*Article*

# Antibiofilm and Enzyme Inhibitory Potentials of Two Annonaceous Food Spices, African Pepper (*Xylopia aethiopica*) and African Nutmeg (*Monodora myristica*)

**Alfred Ngenge Tamfu** [1,2,3,*], **Ozgur Ceylan** [2], **Selcuk Kucukaydin** [4], **Mehmet Ozturk** [5], **Mehmet Emin Duru** [5] **and Rodica Mihaela Dinica** [3,*]

[1] Department of Chemical Engineering, School of Chemical Engineering and Mineral Industries, University of Ngaoundere, Ngaoundere 454, Cameroon

[2] Food Quality Control and Analysis Program, Ula Ali Kocman Vocational School, Mugla Sitki Kocman University, 48147 Ula Mugla, Turkey; ozgurceylan@mu.edu.tr

[3] Department of Chemistry Physical and Environment, Faculty of Sciences and Environment, "Dunarea de Jos" University of Galati, 111 Domneasca Street, 800201 Galati, Romania

[4] Department of Medical Services and Techniques, Koycegiz Vocational School of Health Services, Mugla Sitki Kocman University, 48800 Mugla, Turkey; selcukkucukaydin@gmail.com

[5] Department of Chemistry, Faculty of Science, Mugla Sitki Kocman University, 48000 Mugla, Turkey; mehmetozturk@mu.edu.tr (M.O.); eminduru@mu.edu.tr (M.E.D.)

[*] Correspondence: macntamfu@yahoo.co.uk (A.N.T.); rodica.dinica@ugal.ro (R.M.D.); Tel.: +237-675590353 (A.N.T.); +33-6130-251 (R.M.D.)

Received: 29 October 2020; Accepted: 27 November 2020; Published: 29 November 2020

**Abstract:** Food pathogens represent an important health threat, and it is relevant to study the effect of foodstuffs such as spices which can inhibit bacterial growth. This study reports the antimicrobial, antibiofilm, and enzyme (Acetylcholinesterase, Butyrylcholinesterase, urease, tyrosinase) inhibitory activities of two medicinal food spices belonging to the Annonaceae family, *Monodora myristica* and *Xylopia aethiopica*. GC-MS (gas chromatography mass spectrometry) analysis of silylated samples of Methanol-Dicloromethane (50:50) extracts of both plants led to the identification of nine compounds in *M. myristica* and seven compounds in *X. aethiopica*. *M. myristica* and *X. aethiopica* had the same minimum inhibitory concentration (MIC) values of 0.625 mg/mL and 2.5 mg/mL on *C. albicans* and *E. coli*, respectively. However, *M. myristica* had better activity than *X. aethiopica* on *Staphylococcus aureus*, while *Pseudomonas aeruginosa* was more susceptible to *X. aethiopica* than *M. myristica*. The lowest MIC value was 0.1325 mg/mL, exhibited by *M. myristica* on *S. aureus*. Both extracts showed good antibiofilm activity. On *S. aureus*, at the same concentration, *M. myristica* had better antibiofilm activity than *X. aethiopica*. On *E. coli* and *Candida albicans*, *X. aethiopica* had better antibiofilm activity than *M. myristica* at the same concentration. *X. aethiopica* showed better violacein inhibition in *Chromobacterium violaceum* CV12472, as its percentage inhibition of violacein varied from 80.5% ± 3.0% at MIC to 5.6 ± 0.2 at MIC/8, as compared to *M. myristica* with 75.1% ± 2.5% at MIC and 15.5% ± 1.1% at MIC/8. The anti-motility activity by swimming and swarming inhibition on *P. aeruginosa* PA01 was low at test concentrations and in both models, *M. myristica* showed higher motility inhibition than *X. aethiopica*. Although in enzyme inhibitory assays all extracts had low inhibitions compared to standards tested at the same concentrations, the results show that these plants can be used to manage food-borne infections.

**Keywords:** African food spices; GC-MS (gas chromatography mass spectrometry); antimicrobial; antibiofilm; violacein inhibition; swarming inhibition; swimming inhibition; anticholinesterase; antiurease; antityrosinase

---

## 1. Introduction

In every region of the world, selected indigenous plants are used as foods and spices, and investigating their chemical composition and bioactivities has become an interesting field of research. This is because it reveals both the nutritive value and medicinal potential of these food materials used as remedies especially for recalcitrant infectious diseases resulting from food contamination. Spices are defined by Corn et al. (1999) as ingredients usually from vegetables or different dried plant parts like barks, seeds, and leaves added in nutritionally small quantities to food in order to improve its color, taste, or flavor and may also play the role of preservatives that inhibit harmful bacterial growth [1,2]. Most spices are added to food recipes primarily for flavoring, seasoning, and imparting aroma to foods rather than for their nutritional benefits, and for this reason, their phytochemical compositions as well as bioactivities are still under-studied [3]. There is need for the search of locally available nutritional food excipients in Africa and other low income countries which are plagued with poor nutrition [4]. *Xylopia aethiopica* and *Monodora myristica* are two widely consumed food spices in Africa and both belong to the Annonaceae family of plants and have proven to be potent in managing microbial and fungal infections [5]. Annonaceous plants are commonly called custard apples and comprises over 2300 species which possess good biological activities [6].

*X. aethiopica*, also called African pepper, is a highly consumed food spice in Africa and it is used traditionally to manage rheumatism, bronchitis, headache, asthma, stomach-aches, neuralgia, dysenteric conditions, wounds and sores, constipation, epilepsy, fertility, and the ease of childbirth [7]. It has been shown to possess anticancer, antidiabetic, antimalarial, antioxidant, enzyme inhibitory, antimicrobial, and antibacterial properties and also protects against liver and kidney damage [3,7–14].

*Monodora myristica*, also called African nutmeg, bears many fruits. The seeds of this plant are mostly used in dry powdered form which is used as spice in desserts, stews, soups, and cakes. It is also marketed as whole seeds. These plants' seeds and powders are used in repelling insects, as a stimulant, and as remedy for sores, headache, and stomach disorders. In medicine, the bark is used in treatments of stomachaches, febrile pains, eye diseases, and hemorrhoids [15,16]. This plant possesses antioxidant, antisplasmodic, antiulcer, antimicrobial, cytotoxic, antinociceptive, anti-inflammatory, anticancer, and hepatoprotective activities, and it has been proven in male Wistar rats with induced hypercholesterolemic to modulate lipid peroxidation and bring down cholesterol levels [3,17–25].

These plants have been reported to possess antimicrobial activities, with more emphasis on their essential oil and little information about their extracts, meanwhile the plant seeds are consumed in whole. No studies have reported the ability of these plants to inhibit quorum-sensing-mediated effects in bacteria such as biofilm, violacein production, and swarming and swimming motilities. The aim of this work is to evaluate the ability of *M. myristica* and *X. aethiopica* extracts to inhibit biofilm formation, violacein production, and swimming and swarming motilities, as well as acetylcholinesterase, butyrylcholinesterase, urease, and tyrosinase enzymes.

## 2. Materials and Methods

### 2.1. Plant Material and Extraction

The seeds of *Xylopia aethiopica* and *Monodora myristica* were purchased from the Bamenda food market. Both spicy plants were identified by Mr. T. Fulbert, a botanist working at the National Herbarium of Cameroon and compared with existing voucher specimens 28725/SRF Cam. (*X. aethiopica*) and 49544/HNC (*M. myristica*). Of each of the seeds, 200 g were powdered and subjected to maceration extraction. Then, 2 L of dichloromethane/methanol mixture in the ratio 1:1 was used as the solvent for extraction process and the mixture was allowed to stand at room temperature for 48 h with intermittent stirring. After this, the supernatant was carefully decanted and filtered using a Whatman number 1 filter paper. This filtrate was evaporated on a Rotary evaporator to remove the solvent. This process was repeated three times for each sample to yield crude extracts of *X. aethiopica* (27 g) and *M. myristica* (41 g).

### 2.2. GC-FID (Gas Chromatography Flame Ionization Detector) and GC-MS Analyses

Prior to the GC-MS analyses, the samples were silylated using BSTFA [bis (trimethylsilyl)-trifluoroacetamide], according to the method described by Talla and coworkers [26]. GC-FID and GC-MS were performed as described elsewhere [27]. The GC-MS profiles of the extracts were achieved on a gas chromatograph (Hewlett-Packard 5890, Bunker Lake Blvd, Ramsey, MN, USA) with a JEOL MS-600H mass spectrometer (Tokyo, Japan) as detector. Prior to this, GC-FID was performed on a Shimadzu GC-17(Shimadzu Corp., Kyoto, Japan). We used helium (1 mL/min) as the carrier gas at a split ratio of 1:10 in a SPB-5VR capillary column of length 30 m and inner diameter 0.25 mm. The initial temperature of the oven was varied as follows: 60 °C for 3 min, increased at the rate of 5 °C/min to 180 °C, and finally at 7 °C/min to 300 °C final temperature. A ZB-5MSVR column of 30 m length and 0.25 mm inner diameter was used for GC-MS and the same temperature conditions as for GC-FID were applied. Next, 250 °C and 70 eV was applied for the ion source. Mass spectral fingerprints were used for identification on the NIST library and compared with some data reported.

### 2.3. Determination of MIC (Minimal Inhibitory Concentration)

The bacterial and fungal strains *Staphylococcus aureus* (ATCC 25923), *Escherichia coli* (ATCC 25922), *Candida albicans* (ATCC 10239), *Chromobacterium violaceum* (CV12472), and *Pseudomonas aeroginosa* (PA01) were used.

The broth dilution method, as described by the CLSI (Clinical and Laboratory Standards Institute, 2006), was applied to determine the MIC values [28]. The lowest concentration of extract at which no bacterial growth was visible was considered as the MIC. Mueller-Hinton Broth (MHB) was used as the medium and the bacterial concentration used had a density of $5 \times 10^5$ colony-forming units (CFU)/mL. Into 96-well microtiter plates, containing extracts at concentrations (10, 5, 2.5, 1.25, 0.625, 0.312 mg/mL), 100 µL of microbial cell solutions were inoculated and incubated for 24 h at 37 °C, after which the MICs were determined and recorded.

### 2.4. Assay of Inhibition of Bacterial Biofilm Formation by Extracts

The ability of the extracts to prevent biofilm formation by bacteria (*S. aureus*, *E. coli* and *C. albicans*) was evaluated at concentrations of 1, 1/2, 1/4, 1/8, and 1/16 (MIC) using the microplate antibiofilm method [29]. Tryptose-Soy Broth (TSB) containing 0.25% glucose were filled into wells with or without extracts, and 200 µL of 1% overnight bacterial cell cultures ($5 \times 10^5$ CFU/mL) were added and incubated at 37 °C for 48 h. The negative control wells contained only TSB and bacterial cells. The planktonic bacteria were washed after incubation, and the remaining bacteria were stained with 0.1% crystal violet solution and allowed to sit for 10 min. The dye was washed out carefully with distilled water, after which 200 µL of ethanol or 33% glacial acetic were filled into the wells of the microplates. Next, 125 µL of the resulting solution were transferred using a pipette into sterile tubes and the total volume was made up to 1 mL by adding distilled water. The optical density of each tube was read at 550 nm, and the equation below was used to calculate the percentage inhibition of the biofilm formation. Each experiment was done three times.

$$\text{Biofilm inhibition } (\%) = \frac{OD\ 550\ control - OD\ 550\ sample}{OD\ 550\ control} \times 100$$

### 2.5. Violacein Pigment Inhibition Assay

This assay measures the qualitative quorum-sensing potential of the extracts using *Chromobacterium violaceum* ATCC 12472 strain [30]. *C. violaceum* CV12472 was grown overnight and 10 µL of it were put into microtiter sterile plates filled with 200 µL of LB broth. Sub-MIC concentrations of extracts were added and then incubated at 30 °C for 24 h. Control plates contained only LB broth and *C. violaceum* ATCC 12472. The decrease in the production of violacein pigment was measured by taking the absorbance at 585 nm. The violacein inhibition percentage of the extracts were calculated as follows:

$$\text{Violacein pigment inhibition } (\%) \;=\; \frac{OD\;585\;control - OD\;585\;sample}{OD\;585\;control} \times 100$$

### 2.6. Swimming and Swarming Motility Inhibition on Pseudomonas Aeruginosa PA01

The ability of the extracts to inhibit swarming motility in *P. aeruginosa* PA01 was performed as described elsewhere [30]. Briefly, swarming plates consisting of, 0.5% NaCl, 1% peptone, 0.5% agar, and 0.5% D-glucose together with extracts at three concentrations of 50, 75, and 100 μg/mL were prepared. *P. aeruginosa* PAO1 was grown overnight and 5 μL of it were point-inoculated at the center, and the swarm plates were wrapped with paraffin and inoculated in an upright position for 18 h. The plates not containing extracts were used as controls. The swarming movement was measured from the swarm diameter fronts.

The swimming plates consisted of 1.5% agar, 1% peptone, 0.5% NaCl, and 0.5% D-glucose together with the extracts at 50, 75, and 100 μg/mL concentrations. The same bacteria *P. aeruginosa* PAO1 was inoculated as in the swarming model. The inoculation was done for 18 h and control plates did not contain extracts. Swim zone diameter for samples and controls were used in calculating the percentage inhibition of swimming motility.

### 2.7. Cholinesterase Inhibition Assay

Ellman's Method was used to evaluate the acetylcholinesterase (AChE) and butyrylcholinesterase (BChE) inhibition potential of extracts, and slight modifications were made [31]. Next, 96-well microplates were used and acetylthiocholine iodide (0.71 mM) was the substrate in AChE assay, while butyrylthiocholine chloride (0.2 mM) was the substrate for the BChE assay. In a 96-well plate containing 150 μL sodium phosphate buffer 100 mM (pH = 8), 10 μL of the sample were mixed with 20 μL of enzymes AChE or BChE. The mixture was incubated at 25 °C for 15 min, after which 10 μL of substrates and 10 μL of Ellman's Reagent (DTNB 0.5 mM) were added and the volume made up to 200 μL. The absorbance was measured at 412 nm for 10 min. The percentage inhibition of AChE or BChE was determined control using the formula:

$$(E - S)/E \times 100 \tag{1}$$

where:

E: activity of enzyme with control.
S: activity of enzyme with sample.

The experiments were repeated three times. Galantamine was used as the standard.

### 2.8. Tyrosinase Inhibition Assay

A spectrophotometrical method was used to evaluate the anti-tyrosinase activity of extracts in which tyrosinase enzyme from mushrooms was used following a method described elsewhere with a slight modification [32]. The substrate used in this assay was L-Dopa, while kojic acid was used as the standard inhibitor of tyrosinase. The percent inhibition of the enzyme (Inhibition %) by the extracts was calculated at each sample concentration (μg/mL) in a similar manner as in the AChE and BChE assay.

### 2.9. Urease Inhibition Assay

The indophenol method, in which the production of ammonia is measured, was used to determine the potential of extracts to inhibit urease [33]. A mixture of 25 μL of a Jack bean source urease enzyme, phosphate buffer 100 mM (pH 8.2), and 50 μL of urea 100 mM was prepared and after the adding of the samples (10 μL, 1 mM), it was incubated for 15 min at 30 °C. Subsequently, 45 μL of phenol reagent 1% (*w/v*) and 70 μL of 0.005% (*w/v*) alkali reagent were both added into each well and the mixture further incubated for 50 min. The standard used was Thiourea. The absorbances were recorded at 630 nm and the % inhibitions were calculated.

*2.10. Statistical Analysis*

Each activity was done in triplicate. The results were recorded as the means ± standard error of the mean. Fisher's test was used to determine the significant differences between means; $p < 0.05$ were regarded as significant.

## 3. Results

*3.1. GC-MS Chemical Composition*

Nine compounds (Figure 1), including 1-monolinoleoylglycerol, 6,9,12-octadecatrienoic acid benzyl ester, 3-hydroxyspirost-8-en-11-one, ethyl-3,4,5-trimethoxybenzoate, palmitic (hexadecanoic) acid, ursodeoxycholic acid and sugars such as glycerol, glucose, and sucrose were identified in *M. myristica*. On the other hand, seven compounds (Figure 2), were identified in *X. aethiopica*, these were 3-carene, eucalyptol, 2-hydroxy-4-methylbenzoic acid, abietic acid, 3,21-dihydroxypregnan-4-one alongside two sugars, fructose, and glucose.

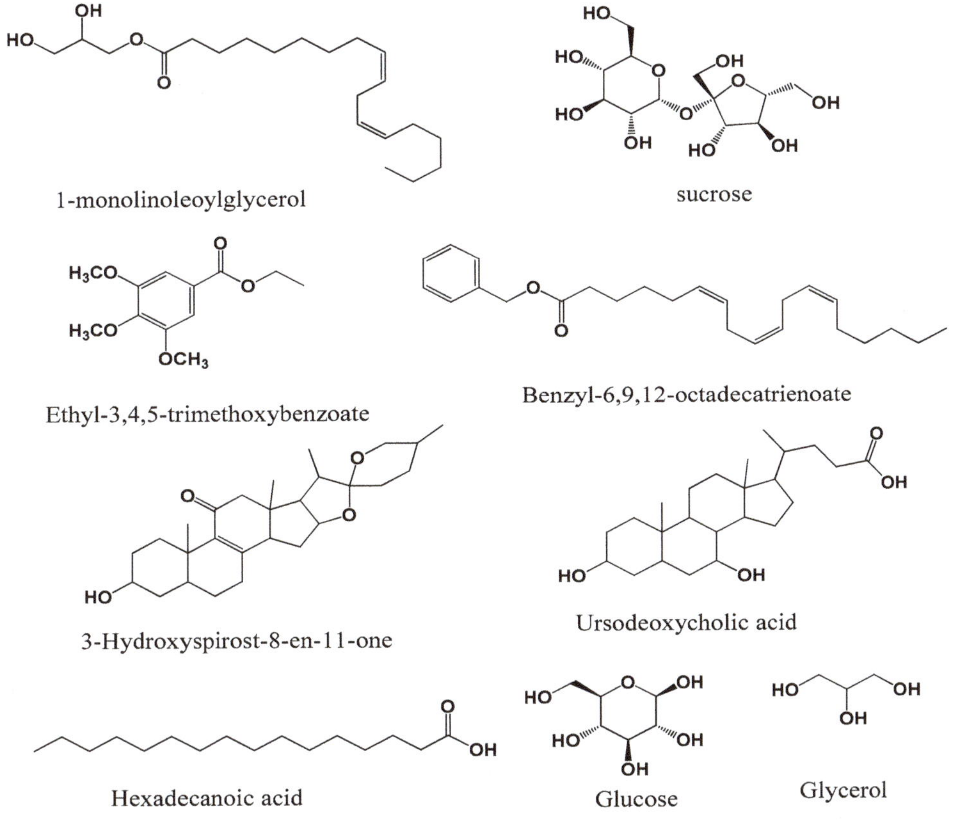

**Figure 1.** Chemical compounds identified in *M. myristica* seed extract by gas chromatography mass spectrometry (GC-MS).

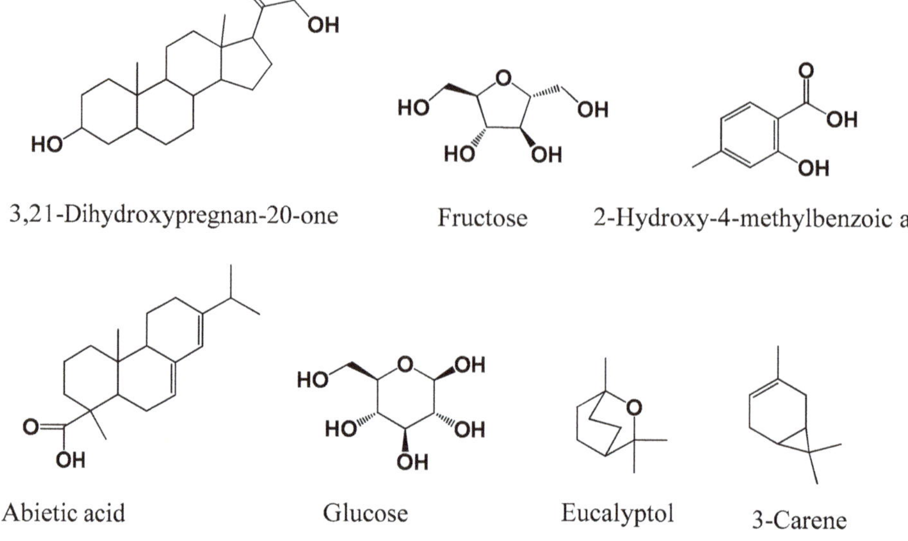

3,21-Dihydroxypregnan-20-one  Fructose  2-Hydroxy-4-methylbenzoic acid

Abietic acid  Glucose  Eucalyptol  3-Carene

**Figure 2.** Chemical compounds identified in *X. aethiopica* seed extract by GC-MS.

### 3.2. Minimal Inhibitory Concentrations

The minimal inhibitory concentrations (MICs) of both extracts are reported in Table 1. Both plants *X. aethiopica* and *M. myristica* had the same MIC values of 0.625 mg/mL and 2.5 mg/mL on *C. albicans* and *E. coli*, respectively. However, *M. myristica* had better activity than *X. aethiopica* on *S. aureus*, while *P. aeruginosa* was more susceptible to *X. aethiopica* than *M. myristica*. The lowest MIC value was 0.1325 mg/mL, exhibited by *M. myristica* on *S. aureus*.

**Table 1.** Antimicrobial activities of *X. aethiopica* and *M. myristica* (minimum inhibitory concentration (MIC) values in mg/mL).

| Microorganisms | Extracts | |
|---|---|---|
| | *X. aethiopica* | *M. myristica* |
| *C. albicans* ATCC 10239 | 0.625 | 0.625 |
| *S. aureus* ATCC 25923 | 0.625 | 0.3125 |
| *E. coli* ATCC 25922 | 2.5 | 2.5 |
| *C. violaceum* CV12472 | 2.5 | 1.25 |
| *P. aeruginosa* PA01 | 0.625 | 1.25 |

### 3.3. Percentage Biofilm Inhibition

Prior to antibiofilm assay, the MIC values were determined and the biofilm inhibition potential of both extracts determined at MIC and sub-MIC concentrations and reported in Table 2. Both extracts had good antibiofilm activity on *S. aureus*. On *S. aureus*, the biofilm inhibition of *X. aethiopica* varied from 45.3% ± 1.5% at MIC to 5.2% ± 1.0% at MIC/8 while that of *M. myristica* varied from 52.6% ± 3.3% at MIC to 11.3% ± 1.5% at MIC/8. On *S. aureus*, at the same concentration, *M. myristica* had better antibiofilm activity than *X. aethiopica*. On *E. coli* and *C. albicans*, *X. aethiopica* had better antibiofilm than *M. myristica* at the same concentration ($p < 0.01$).

**Table 2.** Effects of different concentrations of *X. aethiopica* and *M. myristica* extracts on tested bacteria biofilm formations.

| Strains | Conc. | *X. aethiopica* | *M. myristica* |
|---|---|---|---|
| | | **% of Biofilm Inhibition** | |
| S. aureus ATCC 25923 | MIC | 45.3 ± 1.5 * | 52.6 ± 3.3 |
| | MIC/2 | 30.0 ± 3.4 | 39.0 ± 1.4 |
| | MIC/4 | 16.7 ± 0.4 | 23.8 ± 0.6 |
| | MIC/8 | 5.2 ± 1.0 | 11.3 ± 1.5 |
| | MIC/16 | - | - |
| E. coli ATCC 25922 | MIC | 21.0 ± 2.5 | 18.6 ± 0.5 |
| | MIC/2 | 10.5 ± 0.5 | 8.3 ± 2.0 |
| | MIC/4 | - | - |
| | MIC/8 | - | - |
| | MIC/16 | - | - |
| C. albicans ATCC 10239 | MIC | 31.7 ± 1.5 | 29.6 ± 3.3 |
| | MIC/2 | 22.4 ± 1.3 | 15.3 ± 0.1 |
| | MIC/4 | 09.2 ± 0.5 | 8.9 ± 2.0 |
| | MIC/8 | - | - |
| | MIC/16 | - | - |

*: Data are the mean of three replicates ± SD. Statistically different ($p < 0.01$).

### 3.4. Violacein and Motility (Swimming and Swarming) Inhibition Percentages

The percentage inhibition of violacein pigment synthesis by *C. violaceum* CV12472 was evaluated at MIC and sub-MIC concentrations and reported in Table 3. *X. aethiopica* inhibited violacein production better than *M. myristica*, as it percentage inhibition of violacein varied from 80.5% ± 3.0% at MIC to 5.6 ± 0.2 at MIC/8 as compared to *M. myristica* with 75.1% ± 2.5% at MIC and 15.5% ± 1.1% at MIC/4. At MIC/16, no inhibition was observed.

The swimming and swarming inhibition assay was carried on *P. aeruginosa* PA01 at 100 µg/mL, 75 µg/mL, and 50 µg/mL, and the results are given in Table 3. The anti-motility activity of these plant extracts were low at test concentrations. In both models, *M. myristica* showed higher motility inhibition than *X. aethiopica* ($p < 0.05$).

**Table 3.** Effects of *X. aethiopica* and *M. myristica* extracts at different concentrations on qualitative violacein inhibition and swarming/swimming motility.

| | *X. aethiopica* | | *M. myristica* | |
|---|---|---|---|---|
| Conc. | % of Violacein Inhibition (on *C. violaceum* CV12472) | | | |
| MIC | 80.5 ± 3.0 * | | 75.1 ± 2.5 | |
| MIC/2 | 37.3 ± 0.7 | | 35.8 ± 0.6 | |
| MIC/4 | 19.4 ± 1.0 | | 15.5 ± 1.1 | |
| MIC/8 | 5.6 ± 0.2 | | - | |
| MIC/16 | - | | - | |
| | % motility inhibition (on *P. aeruginosa* PA01) | | | |
| Conc. (µg/mL) | Swimming inh. | Swarming inh. | Swimming inh. | Swarming inh. |
| 100 | 25.1 ± 0.5 | 28.9 ± 2.0 | 29.2 ± 0.0 | 31.7 ± 1.2 |
| 75 | 13.6 ± 0.5 | 12.0 ± 0.5 | 16.5 ± 0.8 | 22.3 ± 1.0 |
| 50 | - | - | - | 11.6 ± 0.5 |

*: Data are the mean of three replicates ± SD. Statistically different ($p < 0.05$).

*3.5. Enzyme Inhibitory Activities*

The capacity of *X. aethiopica* and *M. myristica* to inhibit some selected enzymes acetylcholinesterase, butyrylcholinesterase, tyrosinase, and urease are evaluated and reported in Table 4. On AChE, the activities of *X. aethiopica* (7.80% ± 0.36%) and *M. myristica* (9.65% ± 0.11%) were almost the same but far lower than that of the standard galantamine (80.76% ± 0.52%). This was the same effect for BChE on which percentage inhibitions of *X. aethiopica* and *M. myristica* were 15.48 ± 0.57 and 13.64 ± 0.88 respectively against 74.43% ± 0.30% for galantamine.

*X. aethiopica* (30.45% ± 0.84%) had better inhibition on tyrosinase than *M. myristica* (23.44% ± 1.27%) while on urease, *M. myristica* (15.18% ± 0.86%) was more active than *X. aethiopica* (5.69% ± 0.33%).

**Table 4.** Anticholinesterase, antityrosinase, and antiurease activities (%inh.) of test samples at 100 µg/mL.

| Sample | AChE | BChE | Tyrosinase | Urease |
|---|---|---|---|---|
| | % Inhibition | | | |
| *X. aethiopica* | 7.80 ± 0.36 | 15.48 ± 0.57 | 30.45 ± 0.84 | 5.69 ± 0.33 |
| *M. myristica* | 9.65 ± 0.11 | 13.64 ± 0.88 | 23.44 ± 1.27 | 15.18 ± 0.86 |
| Galantamine | 80.76 ± 0.52 | 74.43 ± 0.30 | NT | NT |
| Kojic acid | NT | NT | NT | 67.58 ± 0.23 |
| Thiourea | NT | NT | 75.17 ± 0.18 | NT |

NT: Not tested. Statistically different ($p < 0.01$).

## 4. Discussion

The chemical composition of both plant extracts were effected using GC-MS. The GC-MS was done after the samples were silylated to enable the detection of a large number of compounds including polar ones. Silylation helps to reduce polarity of compounds contained in the extract while equally rendering them more volatile such that they become detectable on GC-MS. These enabled the identification of major compounds in both extracts. These two plants investigated in this study have evidently been proven to possess antimicrobial activity and have been used extensively to manage infectious diseases [34], but much of this has been attributed to its essential oils [5] with little attention on the extracts. In this study, the extracts have also demonstrated appreciable antimicrobial activity, and this is advantageous, since the plants are consumed principally in crude form. Antimicrobial activity was displayed by both plant extracts and these results corroborate with some findings in which these plant extracts showed antimicrobial activity against some pathogens [35]. However, microbial resistance arises due to quorum-sensing-mediated traits of pathogens which determines the severity of infections as well. The assays that involved quorum-sensing-mediated processes in bacteria are usually focused biofilm formation, violacein pigment production, and swimming and swarming motilities. Prior to these assays, MIC values are determined and they are performed at MIC and sub-MIC concentrations. Thus, working at sub-MIC concentrations, the hypothesis of bactericidal effect of extracts that occurs at high concentrations is eliminated, giving way for QS investigation.

In order to protect themselves during adverse conditions such as immunological defense systems of host, antibiotics, and starvation, many fungi and bacteria cells constitute self-organized and three-dimensional communities in which they will live. These constituted and protected communities are called biofilms and they contribute to the severity of chronic infections as well as the persistence of resistance to drugs and antibiotics [6,36]. Therefore, most antimicrobial agents treat symptoms of planktonic bacteria, and disease will resurface, due to the bacteria which were within biofilms and which will break out when the threat from the antibiotic is over. The potential of these plant extracts to inhibit biofilm formation by test bacteria is a very desirable effect. This can subsequently eliminate bacteria resistance or reduce the severity of microbial infections.

Violacein (violet color) production is one of the quorum-sensing-mediated traits of pathogenic bacteria *C. violaceum* that has received maximum attention and has the advantage that it is easily measurable and quantifiable. Violacein is a biomolecule that has antimicrobial activity against other microorganisms and also improves the *C. violaceum's* ability to cope with environmental stress [37,38]. *C. violaceum* has been highly applied in evaluating the antiinfective potential of many natural and synthetic products, and those with violacein inhibition find application potential alternatives to conventional bactericidal antibiotics.

The bacterial communities that are formed on surfaces constitute a serious health threat and potent contamination. The colonization of various surfaces by bacteria through bacterial motility can be reduced or limited by the use of natural substances and compounds which need to be identified and evaluated for their applications to impede bacterial colonizations [39]. Before forming biofilms, bacteria move by swimming and swarming and colonize surfaces, and this step is believed to be an initial step for the formation of the biofilm in flagellated *P. aeruginosa* PA01. This bacterium can use its flagella to swim towards surfaces followed by swarming and twitching motilities, which enables it to colonize surfaces and establish biofilm communities.

Cholinesterase inhibitors are a group of medicines that block the normal breakdown of acetylcholine. Deterioration of the central nervous system and Alzheimer's disease (AD) result from cholinergic deficiency and are associated with dementia. As a remedy to this situation, various cholinesterase inhibitors such as natural compounds and extracts and synthetic analogues and their hybrids are being employed [40]. Natural medicines are gaining ground because of their low toxicities and few side effects compared to synthetic ones. The enzyme-mediated browning reaction in harvested fruits is usually initiated by the copper-containing oxidase enzyme called tyrosinase. It is an enzyme with multiple functions and is responsible for the undesirable over-pigmentation of human skin in a process that is similar to that which takes place during the browning of fruits. The first two steps in mammalian melanogenesis are catalyzed by this enzyme, and hence the search for potent tyrosinase inhibitors, especially those that can be used safely in cosmetics and foods, is an attraction for researchers [41]. Many bacteria produce the nickel-dependent urease (ureolytic bacteria) enzyme, which is capable of hydrolyzing urea to produce ammonia with the emission of carbon dioxide as well. The emission of ammonia from agriculture is usually associated with these bacteria, and it also causes a number of infectious diseases [42]. Inhibiting urease is beneficial to human health and some notorious bacteria depend on urease-mediated processes for survival. Although in the enzyme inhibitory assays no extract had an inhibition close to that of the standards, these results are moderate because the extracts were tested at the same concentrations as the pure standard compounds.

Many of the compounds identified in these plants have previously been described in some medicinal plant extracts. 1-monolinoleoylglycerol, which is found in *M. myristica* has been described in the extract of *Datura stramonium* and *Salvadora persica* and the extracts of these plants have shown antimicrobial and anticancer activities [43,44]. Certain sugars were detected in both *M. myristica* and *X. aethiopica* plant extracts notably sucrose, glucose, glycerol, and fructose. These sugars were shown to possess antibacterial activities [45,46]. Ethyl-3,4,5-trimethoxybenzoate detected in *M. myristica* and other derivatives of 3,4,5-trimethoxybenzoic acid have been identified in medicinal plant *Mitracarpus scaber* and have demonstrated antimicrobial properties [47], and *Anarcadium occidentale* (cashew gum) also has been synthesized and evaluated for antioxidant and enzyme inhibition [48]. 6,9,12-octadecatrienoic acid phenylmethyl ester was found in the extract of *M. myristica*, and this compound was detected in *Croton bonplandianum* extract which possesses anti-inflammatory and enzyme inhibitory potentials [49]. 3-hydroxyspirost-8-ene-11-one, which was detected in *M. myristica*, has been previously described in *Artemisia annua*, and this plant extract studied showed antioxidant, antimicrobial, and anti-inflammatory activities [50]. Ursodeoxycholic acid was identified in *M. myristica*, and this compound has been synthesized from a natural source [51]. Hexadecanoic acid contained in *M. myristica* has been described in many plants and this compound has good medicinal properties including anti-inflammatory, antioxidant, and enzyme inhibitory activities including anticholinesterase activity [52–54]. The compound 2-Hydroxy-4-methylbenzoic acid present in *X. aethiopica* has been detected previously in *Hemidesmus indicus* and mangrove fungus and has displayed anti-inflammatory, antipyretic, and antioxidant activities [55,56]. In the extract of *X. aethiopica*, abietic acid was identified, and this compound has been detected in *Isodon wightii*, which displayed antibacterial, antioxidant, and anticholinesterase activities [57]. A major component of eucalyptus called eucalyptol or 1,8-cineole identified in *X, aethiopica* extract has been detected in many species of eucalyptus, whose extracts possess antimicrobial and antiviral activities [58,59]. The volatile compound 3-carene was detected in the extract of *X. aethiopica* and has been described as an antimicrobial compound [60]. From these analyses of previous studies, it can be understood that the antimicrobial, antibiofilm, and enzyme inhibitory activities of these extracts of *M. myristica* and *X. aethiopica* can be attributed to these chemical compounds that they contain.

## 5. Conclusions

The investigation of bioactivities of food extracts and food-derived substances is attracting much attention because of their considerable and known safety. These food substances are exploited for their medicinal properties besides their nutritive values especially to combat food-borne infections resulting from various food pathogens. The emergence and spread of drug-resistant strains nowadays threatens the use of conventional antibiotics in order to treat bacterial infections and diseases. New strategies such as biofilm inhibition and quorum-sensing disruption are therefore necessary to overcome persistent infections especially those that involve resistant biofilm formation by the pathogens. Both food spices *X. aethiopica* and *M. myristica* have shown biofilm inhibitory potentials and also violacein inhibition and swimming/swarming motilities inhibitions. These traits are quorum-sensing-mediated processes which help to increase the resistance of bacteria and virulence of their infections. Both extracts displayed low to moderate anticholinesterase activities and tyrosinase and urease inhibitory potentials, which is interesting for these natural food spices. As an added value to the antimicrobial properties of these two natural food spices, they can be desirable as enzyme inhibitors which may present no or milder side effects than conventional synthetic drugs used for this purpose. The bioactivities of these extracts opens a space for the search of the bioactive molecules contained in them.

**Author Contributions:** Conceptualization, A.N.T. and O.C.; methodology, A.N.T., S.K., O.C., R.M.D., M.O. and M.E.D.; investigation, A.N.T., S.K. and O.C.; writing—original draft preparation, A.N.T. and O.C.; writing—review and editing, A.N.T., O.C. and R.M.D.; supervision, O.C., R.M.D., M.O. and M.E.D. Project administration, R.M.D.; funding acquisition, A.N.T. and R.M.D. All authors have read and agreed to the published version of the manuscript.

**Funding:** This research was funded by 'Agence Universitaire de la Francophonie' (AUF) and the Romanian Government through Eugen Ionescu Postdoctoral Mobility Program 2020.

**Acknowledgments:** The authors are grateful to the University of 'Dunarea de Jos' of Galati, Romania and Mugla Sitki Kocman University, Turkey for collaborative material support and for providing a conducive platform for this research.



## References

1. Corn, C. *Scents of Eden: A History of the Spice Trade*; Kindasha: New York, NY, USA, 1999.
2. Chinenye, M.H. Antibacterial Activity of *Piper Guineense*, *Xylopia Aethiopica* and *Allium Cepa* Against Bacteria Isolated from Spoilt Soup Preparations. Ph.D. Thesis, University of Nigeria, Nsukka, Nigeria, 2014; p. 1.
3. Enabulele, S.A.; Oboh, F.O.J.; Uwadiae, E. Antimicrobial, nutritional and phytochemical properties of *Monodora myristica* seeds. *IOSR J. Pharm. Biol. Sci.* **2014**, *9*, 1–6. [CrossRef]
4. Okagu, I.U.; Ogugua, V.N.; Dibor, C.N.I.; Maryann, U.; Nnebe, M.E.; Aniehe, C.C.; Odenigbo, C.J.; Ngwu, U.E. Effects of methanol extraction on some nutritional and antinutrient contents of *Xylopia aethiopica* fruits from Enugu State, Nigeria. *Asian J. Agric. Food Sci.* **2018**, *6*, 84–96. [CrossRef]

5.    Tatsadjieu, L.N.; Essia-Ngang, J.J.; Nyassoum, M.B.; Etoa, F.X. Antibacterial and antifungal activity of *Xylopia aethiopica, Monodora myristica, Xanthoxylum xanthoxyloides* and *Xanthoxylum leprieurii* from Cameroun. *Fitoterapia* **2003**, *74*, 469–472. [CrossRef]

6.    Tamfu, A.N.; Ozgur, C.; Godloves, C.F.; Mehmet, O.; Mehmet, E.D.; Farzana, S. Antibiofilm, antiquorum sensing and antioxidant activity of secondary metabolites from seeds of *Annona senegalensis*, Persoon. *Mic Pathog* **2020**, *144*, 104191. [CrossRef]

7.    Erhirhie, O.; Moke, G.E. *Xylopia Aethiopica*: A Review of its ethnomedicinal, chemical and pharmacological properties earnest. *Am. J. Pharm. Tech. Res.* **2014**, *4*, 22–37.

8.    Choumessi, T.A.; Danel, M.; Chassaing, S.; Truchet, I.; Penlap, V.B.; Pieme, A.C.; Asonganyi, T.; Ducommum, B.; Valette, A. Characterization of the anti-proliferative activity of *Xylopia aethiopica. Cell Div.* **2012**, *7*, 8–10. [CrossRef]

9.    Adefegha, S.A.; Oboh, G. Effect of diets supplemented with Ethiopian pepper [*Xylopia aethiopica* (Dun.) A. Rich (Annonaceae)] and Ashanti pepper [*Piper guineense* Schumach. et Thonn (Piperaceae)] on some biochemical parameters in normal rats. *Asian Pac. J. Trop. Biomed.* **2012**, *16*, S558–S566. [CrossRef]

10.   Soh, D.; Nkwengoua, E.; Ngantchou, I.; Nyasse, B.; Denier, C.; Hannaert, V.; Shaker, K.H.; Schneider, B. Xylopioxyde and other bioactive kaurane-diterpenes from *Xylopia aethiopica* Dunal (Annonaceae). *J. Appl. Pharm. Sci.* **2013**, *3*, 013–019.

11.   Boampong, J.N.; Ameyaw, E.O.; Aboagye, B.; Asare, K.; Kyei, S.; Donfack, J.H.; Woode, E. The curative and prophylactic effects of xylopic acid on *Plasmodium berghei* infection in mice. *J. Parasitol. Res.* **2013**, *2013*, 356107.

12.   Etoundi, C.B.; Kuate, D.; Ngondi, J.L.; Oben, J. Anti-amylase, anti-lipase and antioxidant effects of aqueous extracts of some Cameroonian spices. *J. Nat. Prod.* **2010**, *3*, 165–171.

13.   Abd-Alg, N.N.; El-Kamali, H.H.; Ramadan, M.M.; Ghanem, K.Z.; Farrag, A.R.H. *Xylopia aethiopica* volatile compounds protect against panadol-induced hepatic and renal toxicity in male rats. *World Appl. Sci. J.* **2013**, *27*, 10–22.

14.   Mohammed, A.; Islam, S.M. Antioxidant activity of *xylopia aethiopica* fruits acetone fraction in a type 2 diabetes model of rats. *Biomed. Pharmacother.* **2017**, *96*, 30–36. [CrossRef]

15.   Weiss, E.A. *Spice Crops*; CABI Publishing: Oxon, UK, 2002; pp. 102–103.

16.   Bakarnga-Via, I.; Hzounda, J.B.; Fokou, P.V.; Tchokouaha, L.R.; Gary-Bobo, M.; Gallud, A.; Garcia, M.; Walbadet, L.; Secka, Y.; Dongmo, P.M.; et al. Composition and cytotoxic activity of essential oils from *Xylopia aethiopica* (Dunal) A. Rich, *Xylopia parviflora* (A. Rich) Benth.) and *Monodora myristica* (Gaertn) growing in Chad and Cameroon. *BMC Comp. Altern. Med.* **2014**, *14*, 125. [CrossRef]

17.   Nwozo, S.O.; Fowokemi, K.T.; Oyinloye, B.E. African nutmeg (*Monodora myristica*) lowers cholesterol and modulates lipid peroxidation in experimentally induced hypercholesterolemic male wistar rats. *Int. J. Biomed. Sci.* **2015**, *11*, 86–92.

18.   Ogunmoyole, T.; Inaboya, S.; Makun, J.O.; Kade, I.J. Differential antioxidant properties of ethanol and water soluble phytochemicals of false nutmeg (*Monodora myristica*) seeds. *Int. J. Biochem. Biotechnol.* **2013**, *2*, 253–262.

19.   Lekana-Douki, J.B.; Bongui, J.B.; Oyegue, L.S.L.; Zang-Edou, S.E.; Zatra, R.; Bisvigoua, U.; Druilhe, P.; Lebibi, J.; Ndouo, F.S.T.; Kombila, M. In vitro antiplasmodial activity and cytotoxicity of nine plants traditionally used in Gabon. *J. Ethnopharmacol.* **2011**, *133*, 1103–1108. [CrossRef]

20.   Lacroix, D.; Prado, S.; Kamoga, D.; Kasenene, J.; Namukobe, J.; Krief, S.; Dumontet, V.; Mouray, E.; Bodo, B.; Brunois, F. Antiplasmodial and cytotoxic activities of medicinal plants traditionally used in the village of Kiohima, Uganda. *J. Ethnopharmacol.* **2011**, *133*, 850–855. [CrossRef]

21.   Oyinloye, B.E.; Adenowo, A.F.; Osunsanmi, F.O.; Ogunyinka, B.I.; Nwozo, S.O.; Kappo, A.P. Aqueous extract of *Monodora myristica* ameliorates cadmium-induced hepatotoxicity in male rats. *SpringerPlus* **2016**, *5*, 641. [CrossRef]

22.   Akinwunmi, K.F.; Oyedapo, O.O. In vitro anti-inflammatory evaluation of African nutmeg (*Monodora myristica*) seeds. *Eur. J. Med. Plants.* **2015**, *8*, 167–174. [CrossRef]

23.   Ishola, I.O.; Ikumawoyi, V.O.; AfolayaN, G.O.; Olorife, O.J. Antinociceptive and anti-inflammatory properties of hydroethanolic seed extract of *Monodora myristica* (Annonaceae) in rodents. *West Afr. J. Pharm.* **2016**, *27*, 22–32.

24. Nwankwo, P.O. Comparative study of the antioxidant activities of *Monodora myristica* and *A. sceptrum* on protein and lipid levels of diabetic-induced rats. *IOSR J. Biotechnol. Biochem.* **2015**, *1*, 63–71.

25. Agiriga, A.; Siwela, M. *Monodora myristica* (Gaertn.) Dunal: A plant with multiple food, health and medicinal applications: A review. *Am. J. Food Technol.* **2017**, *12*, 271–284. [CrossRef]

26. Talla, E.; Tamfu, A.N.; Gade, I.S.; Yanda, L.; Mbafor, J.T.; Sophie, L.; Vander-Elst, L.; Popova, M.; Bankova, V. New mono-ether of glycerol and triterpenes with DPPH radical scavenging activity from Cameroonian propolis. *Nat. Prod. Res.* **2017**, *31*, 1379–1389. [CrossRef] [PubMed]

27. Tamfu, N.A.; Jabeen, A.; Tagatsing, F.M.; Tariq, A.B.; Farzana, S. Organic and mineral composition of seeds of *Afrostyrax lepidophyllus* Mildbr. and evaluation of ROS inhibition and cytotoxicity of isolated compounds. *Chem. Afr.* **2019**, *2*, 615–624.

28. CLSI (Clinical Laboratory Standards Institute). *Quality Control Minimal Inhibitory Concentration (MIC) Limits for Broth Dilution and MIC Interpretative Breakpoints (M27-s2)*; Clinical Laboratory of Standards Institute: Wayne, PA, USA, 2006.

29. Merritt, J.H.; Kadouri, D.E.; O'Toole, G.A. *Growing and Analyzing Static Biofilms. Current Protocols in Microbiol*; John Wiley & Sons, Inc.: Hoboken, NJ, USA, 2005.

30. Packiavathy, I.A.S.V.; Agilandeswari, P.; Musthafa, K.S.; Pandian, S.K.; Ravi, A.V. Antibiofilm and quorum sensing inhibitory potential of *Cuminum cyminum* and its secondary metabolite methyl eugenol against Gram negative bacterial pathogens. *Food Res. Int.* **2012**, *45*, 85–92. [CrossRef]

31. Ellman, G.L.; Courtney, K.D.; Andres, V.; Featherstone, R.M. A new and rapid colorimetric determination of acetylcholinesterase activity. *Biochem. Pharmacol.* **1961**, *7*, 88–95. [CrossRef]

32. Khattak, S.; Saeed-Ur-Rehman, S.; Khan, T.; Ahmad, M. In vitro enzyme inhibition activities of crude ethanolic extracts derived from medicinal plants of Pakistan. *Nat. Prod. Res.* **2005**, *19*, 567–571. [CrossRef] [PubMed]

33. Weatherburn, M.W. Phenol-hypochlorite reaction for determination of ammonia. *Anal. Chem.* **1967**, *39*, 971–974. [CrossRef]

34. Kuete, V. Potential of Cameroonian plants and derived products against microbial infections. *Planta Med.* **2010**, *76*, 1479–1491. [CrossRef]

35. Olusimbo, O.A.; Abimbola, R.E.; Chinonye, L.A. Antimicrobial activities of some Nigerian spices on some pathogens. *Agric. Biol. J. N. Am.* **2011**, *2*, 1187–1193.

36. Costerton, J.W.; Stewart, P.S.; Greenberg, E.P. Bacterial biofilms: A common cause of persistent infections. *Science* **1999**, *284*, 1318–1322. [CrossRef] [PubMed]

37. De Vasconcelos, A.T.R.; De Almeida, D.F.; Hungria, M.; Guimarães, C.T.; Antonio, R.V.; Almeida, F.C.; De Almeida, L.G.; De Almeida, R.; Alves-Gomes, J.A.; Andrade, E.M.; et al. The complete genome sequence of *Chromobacterium violaceum* reveals remarkable and exploitable bacterial adaptability. *Proc. Natl. Acad. Sci. USA* **2003**, *100*, 11660–11665.

38. Kothari, V.; Sharma, S.; Padia, D. Recent research advances on *Chromobacterium violaceum*. *Asian Pac. J. Trop. Med.* **2017**, *10*, 744–752. [CrossRef] [PubMed]

39. O'May, C.; Tufenkji, N. The swarming motility of *Pseudomonas aeruginosa* is blocked by cranberry proanthocyanidins and other tannin-containing materials. *Appl. Environ. Microbiol.* **2011**, *77*, 3061–3067. [CrossRef] [PubMed]

40. Sharma, K. Cholinesterase inhibitors as Alzheimer's therapeutics (Review). *Mol. Med. Reports.* **2019**, *20*, 1479–1487. [CrossRef] [PubMed]

41. Chang, T.S. An updated review of tyrosinase inhibitors. *Int. J. Mol. Sci.* **2009**, *10*, 2440–2475. [CrossRef] [PubMed]

42. Svane, S.; Sigurdarson, J.J.; Finkenwirth, F.; Eitinger, T.; Karring, H. Inhibition of urease activity by different compounds provides insight into the modulation and association of bacterial nickel import and ureolysis. *Sci. Rep.* **2020**, *10*, 8503. [CrossRef]

43. Altameme, H.J.; Hameed, I.H.; Kareem, M.A. Analysis of alkaloid phytochemical compounds in the ethanolic extract of *Datura stramonium* and evaluation of antimicrobial activity. *Afr. J. Biotechnol.* **2015**, *14*, 1668–1674.

44. Mohammed, A.B.; Hafiz, A.M.; Hassan, A.A.; Sohier, M.S.; Ashraf, N.A.; Husham, E.H.; Shahnaz, S.; Waquar, A.; Asaad, K. Phytochemical, Cytotoxic, and Antimicrobial Evaluation of the Fruits of Miswak Plant, *Salvadora persica* L. *J. Chem.* **2020**, *2020*, 4521951.

45. Selwyn, S.; Durodie, J. The antimicrobial activity of sugar against pathogens of wounds and other infections of man. In *Properties of Water in Foods*; Simatos, D., Multon, J.L., Eds.; NATO ASI Series (Series E: Applied Sciences); Springer: Dordrecht, The Netherlands, 1985; Volume 90, pp. 293–308.

46. Singh, B.R. Antibacterial Activity of Glycerol, Lactose, Maltose, Mannitol, Raffinose and Xylose. *Med. Noto-Are* **2014**, *2014*, 17223318.

47. Bisignano, G.; Sanogo, R.; Marino, A.; Aquinol, R.; D'Angelo, V.; Germano, M.P.; De-Pasquale, R.; Pizza, C. Antimicrobial activity of *Mitracarpus scaber* extract and isolated constituents. *Lett. Appl. Microbiol.* **2000**, *30*, 105–108. [CrossRef] [PubMed]

48. Katsure, S.V.; Suvarna, A.K.; Dhiraj, M.; Rupali, W.; Mahalaxmi, M.; Sanjay, B.K. Antioxidant and Antiparkinson Activity of Gallic Acid Derivatives. *Pharmacologyonline* **2009**, *1*, 385–395.

49. Vennila, V.; Udayakumar, R. GC-MS Analysis of Leaf, Fruits and Latex of *Croton bonplandianum* Baill. *Int. J. Biochem. Res. Rev.* **2015**, *5*, 187–197. [CrossRef]

50. Hameed, I.H.; Altameme, H.J.; Idan, S.A. *Artemisia annua*: Biochemical products analysis of methanolic aerial parts extract and anti-microbial capacity. *Res. J. Pharm. Biol. Chem. Sci.* **2016**, *7*, 1843–1868.

51. Wang, J.; Gu, X.-Z.; He, L.-M.; Li, C.-C.; Qiu, W.-W. Synthesis of ursodeoxycholic acid from plant-source (20$S$)-21-hydroxy-20-methylpregn-4-en-3-one. *Steroids* **2020**, *157*, 108600. [CrossRef]

52. Zhe, F.; Su, Y.J.; Hyun, A.J.; Jae, S.C.; Byung, S.M.; Mi, H.W. Anticholinesterase and Antioxidant Constituents from *Gloiopeltis furcate*. *Chem. Pharm. Bull.* **2010**, *58*, 1236–1239.

53. Mustapha, N.A.; Runner, R.T.M. GC-MS Analysis and Preliminary Antimicrobial Activity of *Albizia adianthifolia* (Schumach) and *Pterocarpus angolensis* (DC). *Medicines* **2016**, *3*, 3.

54. Krishnan, K.; James, R.F.; Mohan, A. Isolation and characterization of n-hexadecanoic acid from *Canthium parviflorum* leaves. *J. Chem. Pharm. Res.* **2016**, *8*, 614–617.

55. Mahalingam, G.; Krishnan, K. Antidiabetic activity of 2-hydroxy 4-methoxy benzoic acid isolated from the roots of *Hemidesmus indicus* on streptozotocin-induced diabetic rats. *Int. J. Diabetes Metabol.* **2009**, *17*, 53–57.

56. Shao, C.; Guo, Z.; Peng, H.; Peng, G.; Huang, Z.; She, Z.; Lin, Y. A new isoprenyl phenyl ether compound from mangrove fungus. *Chem. Nat. Compd.* **2007**, *43*, 377–380. [CrossRef]

57. Ramnath, M.G.; Thirugnanasampandan, R.; Sadasivam, M.; Mohan, P.S. Antioxidant, antibacterial and antiacetylcholinesterase activities of abietic acid from *Isodon wightii* (Bentham) H. Hara. *Free Rad. Antiox.* **2015**, *5*, 1–5. [CrossRef]

58. Elaissi, A.; Rouis, Z.; Salem, N.A.; Mabrouk, S.; Ben-Salem, Y.; Salah, K.B.; Aouni, M.; Farhat, F.; Chemli, R.; Harzallah-Skhiri, F.; et al. Chemical composition of 8 eucalyptus species' essential oils and the evaluation of their antibacterial, antifungal and antiviral activities. *BMC Complement. Altern. Med.* **2012**, *12*, 81. [CrossRef] [PubMed]

59. Clavijo-Romero, A.; Quintanilla-Carvajal, M.X.; Ruiz, Y. Stability and antimicrobial activity of eucalyptus essential oil emulsions. *Food Sci. Technol. Int.* **2018**, *25*, 24–37. [CrossRef] [PubMed]

60. Huizhen, S.; Haiming, C.; Xiaolong, W.; Yueying, H.; Yonghuan, Y.; Qiuping, Z.; Weijun, C.; Wenxue, C. Antimicrobial Activity and Proposed Action Mechanism of 3-Carene against *Brochothrix thermosphacta* and *Pseudomonas fluorescens*. *Molecules* **2019**, *24*, 3246.

**Publisher's Note:** MDPI stays neutral with regard to jurisdictional claims in published maps and institutional affiliations.

*Article*

# Oregano Phytocomplex Induces Programmed Cell Death in Melanoma Lines via Mitochondria and DNA Damage

**Valentina Nanni** [1,†], **Gabriele Di Marco** [1], **Gianni Sacchetti** [2], **Antonella Canini** [1] and **Angelo Gismondi** [1,*]

1    Department of Biology, University of Rome "Tor Vergata", Via della Ricerca Scientifica 1, 00133 Rome, Italy; nanniva@gmail.com (V.N.); gabriele.di.marco@uniroma2.it (G.D.M.); canini@uniroma2.it (A.C.)

2    Terra&Acqua Tech-Research Unit 7, Pharmaceutical Biology Lab, Department of Life Sciences and Biotechnology, University of Ferrara, Piazzale Luciano Chiappini 3, 44123 Ferrara, Italy; scg@unife.it

*    Correspondence: gismondi@scienze.uniroma2.it; Tel.: +39-06-7259-4333

†    Current address: Department of Molecular Medicine, University of Rome "Sapienza", Viale Regina Elena 291, 00161 Rome, Italy.

Received: 16 September 2020; Accepted: 15 October 2020; Published: 17 October 2020

**Abstract:** Plant secondary metabolites possess chemopreventive and antineoplastic properties, but the lack of information about their exact mechanism of action in mammalian cells hinders the translation of these compounds in suitable therapies. In light of this, firstly, *Origanum vulgare* L. hydroalcoholic extract was chemically characterized by spectrophotometric and chromatographic analyses; then, the molecular bases underlying its antitumor activity on B16-F10 and A375 melanoma cells were investigated. Oregano extract induced oxidative stress and inhibited melanogenesis and tumor cell proliferation, triggering programmed cell death pathways (both apoptosis and necroptosis) through mitochondria and DNA damage. By contrast, oregano extract was safe on healthy tissues, revealing no cytotoxicity and mutagenicity on C2C12 myoblasts, considered as non-tumor proliferating cell model system, and on *Salmonella* strains, by the Ames test. All these data provide scientific evidence about the potential application of this food plant as an anticancer agent in in vivo studies and clinical trials.

**Keywords:** oxidative stress; necroptosis; plant extract; secondary metabolite; γH2AX; copper

---

## 1. Introduction

More than one-third of all pharmaceutical molecules approved by the Food and Drug Administration and by the European Medicines Agency are natural compounds, or their derivatives, and about one-quarter of them specifically originate from plants [1–4]. In detail, over 60% of the anticancer drugs are phytochemicals, such as alkaloids and polyphenols, confirming that the plant kingdom is a valuable source of chemopreventive and chemotherapeutic agents [5–7].

Overall, scientific data have documented that this type of metabolites exerts an inhibitory effect on a broad range of mammalian tumor cell lines in in vitro and in vivo systems [8,9]. According to the literature, the main mammalian cellular and molecular mechanisms influenced by plant molecules are those that involve the following targets: nuclear factor kappa-light-chain-enhancer of activated B cells (NF-κB); protein tyrosine kinases (PTKs); target mitogen-activated protein kinases (MAPKs); cyclooxygenase (COX-2); cyclin-dependent kinases (Cdks); phosphoinositide 3-kinase (PI3K) interactors; activator protein 1 (AP1); and cytoskeleton components [10–22].

Among all biological properties, the free radical scavenging activity seems to be the most validated device employed by phytochemicals to inhibit tumor cell growth, counteracting reactive oxygen species (ROS) production and limiting protein, lipid, and DNA oxidation [23]. Nevertheless, a growing body of evidence brought this opinion into question, arguing that the antiradical activity of the natural compounds has a potential role in chemoprevention, but it cannot fully explain the relative antitumor effect [24,25]. In addition, it is important to underline that a wide number of plant secondary metabolites have also shown unexpected pro-oxidant consequences, including DNA damage and apoptosis, especially at high concentrations and in the presence of transition metal ions [26–29].

*Origanum vulgare* L., also known as oregano, is a Mediterranean plant species belonging to the Lamiaceae family which, nowadays, represents one of the most used culinary herbs. However, the application of oregano in several ethnobotanical practices, including folk medicine, dates to ancient times. Regarding its phytotherapic effect, various investigations have been performed, documenting that oregano essential oil possesses antimicrobial, antiviral, antifungal, antioxidant, anti-inflammatory, digestive, expectorant, neuroprotective, antispasmodic, and antidiabetic properties, simultaneously. Moreover, some literature works have also associated a strong anticancer activity to such type of plant extract. For these reasons, *O. vulgare* is capturing greatly the attention of the food, cosmetic, and pharmaceutical industries [30–35].

According to all this evidence, the present research aimed at investigating the molecular mechanism underlying the antineoplastic effect of *O. vulgare* L. ssp. *hirtum* phytocomplex against murine (B16–F10) and human (A375) melanoma cells.

## 2. Materials and Methods

### 2.1. Plant Material

*Origanum vulgare* L. plants were collected at the *Vatopedi Holy Monastery* on Mount Athos (Greece), in the summer of 2018. The plant material was transferred to the Botanical Garden of Rome "Tor Vergata", where its taxonomic identity was confirmed by Prof. Antonella Canini and Prof. Angelo Gismondi, based on morphological features. A part of the sample was deposited in the *Herbarium* of the Botanical Garden (voucher n. 127C), while the remaining portion was dried out (for 7 days at 37 °C) and used for the present research. In particular, the whole dried plants were powdered in liquid nitrogen, resuspended in 50% ethyl alcohol (200 mg/mL) and incubated, in agitation, for 24 h in the dark. After centrifugation for 20 min at 11,000 g, the supernatant was filtered (0.22 μm), completely desiccated at 30 °C by a vacuum drying system (Concentrator Plus, Eppendorf, Hamburg, Germany), and stored at −80 °C.

### 2.2. Total Phenol and Flavonoid Content

Hydroalcoholic oregano extract (HCOE) was solubilized in 50% ethyl alcohol at the final concentration of 200 mg/mL. The phenolic content in HCOE was measured according to the Folin–Ciocalteu modified method, as described in Impei et al. [36]. Results were reported as μg of gallic acid equivalents per gram of dried plant material (μg GAE/g DMW), applying a gallic acid calibration curve (0–30 mg/L). The amount of flavonoids in OE was assessed by the aluminium chloride colorimetric method [37]. Data were reported as μg of quercetin equivalents per gram of dried plant material (μg QE/g DMW), using a quercetin calibration curve (0–50 mg/L).

### 2.3. High-Performance Liquid Chromatography-Diode Array Detector (HPLC-DAD) and Gas Chromatography-Mass Spectrometry (GC-MS) Analyses

HCOE was characterized by an HPLC system (Shimadzu, Kyoto 604-8511, Japan) associated with an SPD-M20A diode array detector (DAD, Shimadzu, Kyoto 604-8511, Japan) and a Phenomenex Luna C18(2) (3 μm × 4.6 mm × 150 mm) column. A flow of 0.95 mL/min was applied, using formic acid 1% (buffer A) and methanol (buffer B). The following elution gradient was adopted: $t_{0\ min}$ (A 85%, B 15%); $t_{20\ min}$ (A 65%, B 35%); $t_{55\ min}$ (A 10%, B 90%); $t_{68\ min}$ (A 85%, B 15%); $t_{70\ min}$ (A 85%, B 15%). UV–visible absorption spectra at 280 and 340 nm were monitored. Plant metabolites were identified and quantified comparing their retention time, absorbance spectrum, and chromatographic peak area with those of relative pure standards (Sigma-Aldrich) at different concentrations. The amount of each detected molecule was reported as µg per 100 mg of dried plant material (µg/100 mg DMW).

GC-MS analysis was performed exactly as described in Nanni et al. [29]. In particular, to perform this investigation, HCOE was solubilized in 100% methanol at the final concentration of 200 mg/mL and then injected in the instrument.

### 2.4. Cell Cultures and Plant Treatments

Murine melanoma cells (B16–F10), human melanoma cells (A375), and murine myoblasts (C2C12) were cultured under standard conditions [38,39] in Dulbecco's Modified Eagle's Medium (DMEM). For cell experiments, HCOE was solubilized in sterilized PBS 1X, at the concentration of 1200 mg of dried plant material equivalent per mL, and added directly to the cell medium at specific doses. In particular, treatments were performed by exposing cells, for 4, 24 or 48 h, at 2, 4, 6, 8, or 10 mgs of dried plant material equivalent per mL of culture medium. Control cells (CNT) were treated only with PBS, at the highest volume of treatment, to check the influence of this non-toxic solvent on cells.

### 2.5. Cell Proliferation, Selectivity Index, and Cell Cycle Analysis

Cell viability was evaluated by the MTT kit (Sigma-Aldrich Merck, Darmstadt, Germany), as reported in the relative guidelines. Results were reported as percentage variation compared to the control (CNT), which was considered as a unit (100%). Plant extract cytotoxicity was measured using the Trypan Blue (1%, *w/v*) exclusion test and counting dead cells by a Neubauer-modified chamber. The selectivity index (SI) of HCOE on tumor and non-tumor cells was measured according to the following formula: SI = $IC_{50}$ non-tumor cell line/$IC_{50}$ cancer cell line (considering that $IC_{50}$ represented the concentration at which 50% of cell proliferation was inhibited). Cell cycle analysis was performed by a FACSCalibur instrument (Beckton and Dickinson, Le Pont-de-Claix, France) associated to CellQuest software, counting 10,000 events per sample and using the protocol documented in Gismondi et al. [40]. Cytofluorimetric data were shown as a percentage of cells in G0/G1, S, G2/M, and sub-G1 phase. Other treatments were performed by necrostatin-1 (NEC-1; 20 µM, 48 h), Z-VAD-FMK (Z-VAD; 20 µM, 48 h), and Paclitaxel (TAX; 20 nM, 8 h) (Sigma-Aldrich).

### 2.6. Mutagen and Mutagen-Protective Activity

An Ames test was carried out on *Salmonella typhimurium* strains (TA97a, TA98, TA100 and TA1535) in order to evaluate the mutagen and mutagen-protective activity of HCOE. The assays were carried out as widely reported by Rossi et al. [41,42]. In detail, the mutagenic activity was determined by counting *Salmonella* colonies (Colony Counter 560; Suntex Instruments Company Ltd., New Taipei City, Taiwan) in plates treated with different concentrations of oregano extract in the presence and absence of S9 mix metabolic activation. The results were considered positive (potential mutagen) if the amount of revertant colonies was, at least, double that of the negative control. To determine the potential mutagen protection capacity of HCOE (concentration range: 0.01–0.1 mg/plate), bacteria were exposed to mutagenic agents, with or without S9 mix, and exposed or not to different concentrations of HCOE. Used mutagens were 2-nitrofluorene (2 µg/plate; Sigma-Aldrich) for TA97a, TA98, and TA1535; $NaN_3$ (2 µg/plate; Sigma-Aldrich) for TA100 without S9; and 2-aminoanthracene (2 µg/plate; Sigma-Aldrich) for all *Salmonella* strains cultured with S9 mix. Data were expressed as CFU/plate. The inhibition rate (IR) of HCOE for mutagenic induction was measured according to the formula: IR (%) = (A−B) × 100/A (where A and B represent the number of revertants in positive controls or in plates with HCOE, without spontaneous colonies, respectively). Negative controls, represented by dimethylsulfoxide (DMSO) treated strains, were performed to evaluate the background of spontaneous revertants.

### 2.7. Protein Analysis

Cells were lysed in High Salt Buffer (2 mM $CaCl_2$, 350 mM KCl, 50 mM Tris HCl pH 7.4, 1 mM $MgCl_2$) containing 1% protease inhibitor cocktail and 1% NP40. Proteins whose concentration was measured by the Bradford method [43] were separated by SDS-PAGE and transferred onto nitrocellulose membrane. Protein signals were detected by a chemiluminescent kit (Luminol Reagent; Santa Cruz Biotechnology, Dallas, TX, USA) and a VersaDoc Imaging System associated with Quantity One software (Bio-Rad). After normalization with GAPDH, the results were indicated as percentage change compared to the CNT, which was considered as a unit (100%). The antibodies (Santa Cruz Biotechnology) used for Western blotting analyses were as follows: mouse monoclonal GAPDH; mouse monoclonal microphthalmia-associated transcription factor (Mitf); mouse monoclonal p53; mouse monoclonal Parp-1; mouse monoclonal caspase-3 (Casp-3); rabbit polyclonal Bcl-2; rabbit polyclonal Bax; mouse monoclonal control outer mitochondrial membrane protein TOMM20; mouse monoclonal cytochrome c (Cycs); peroxidase-conjugated rabbit, and mouse secondary antibodies. Staurosporine (STS; 2 μM, 4 h) was used as an inducer for apoptosis (Sigma-Aldrich).

### 2.8. Real-Time-PCR (RT-PCR) Assay

Total RNA was extracted by a Pure Link RNA Mini Kit (Invitrogen, Thermo Fisher Scientific, Waltham, MA, USA). RNA concentration and purity were evaluated with a Nanodrop ND1000 spectrophotometer (Thermo Fisher Scientific, Waltham, MA, USA). For cDNA synthesis, 2.5 μg of RNA were incubated for 2 min at 65 °C with 0.4 mM of each dNTP (Euroclone, Milan, Italy). Then, 40 units of RNA inhibitor (Promega, Madison, WI, USA), 0.5 μg random hexamer primers (Invitrogen, Thermo Fisher Scientific, Waltham, MA, USA), 200 units of Moloney murine leukemia virus reverse transcriptase (Promega), 1× enzyme buffer, and 10 mM dithiothreitol were added to reach the final volume of 25 μL. The mix was incubated for 90 min at 37 °C. RT-PCR reactions were carried out in 20 μL of volume composed of 10 ng of cDNA, 5 μM of each primer, and 50% SYBR green (Kapa SYBR Fast qPCR kit; Kapa Biosystems, Roche, Wilmington, MA, USA, Country). cDNA amplification was carried out in an IQ5 thermocycler (Bio-Rad) with the following method: (i) initial denaturation at 95 °C, 4 min; (ii) 45 cycles of denaturation at 95 °C for 20 s (sec), primer annealing at 60 °C (for all genes) for 30 s, and extension at 72 °C for 30 s; and (iii) production of disassociation curve, from 50 to 90 °C (rate: 0.5 °C every 5 s), for the verification of the results. The $2^{-\Delta\Delta Ct}$ formula was used to measure mRNA concentration for each gene: in detail, the threshold cycle (Ct) of the target gene monitored in the treated sample was normalized for the internal reference gene (β-actin, ACTB; ΔCt) and for the respective value of the control sample (ΔΔCt), which was considered as a unit (100%). Supplementary Materials Table S1 reports the list of primers used in this work: microphthalmia-associated transcription factor (MITF), tyrosinase-related protein 1 (TYRP1), tyrosinase (TYR), P21, P27, P53, cyclin-dependent kinase 1 (CDK1), cyclin B1 (CCNB1), and β-actin (ACTB) [44–48].

### 2.9. Reactive Species Level and Mitochondrial Mass and Membrane Potential Measurement

Intracellular reactive oxygen (ROS) and nitrogen (RNS) species, mitochondrial mass, and mitochondrial transmembrane potential were measured by 2′,7′-dichlorodihydrofluorescein diacetate (DCFH-DA; green signal; 10 μM, 15 min), 4-amino-5-methylamino-2′,7′-difluorofluorescein diacetate (DAF-FM DA; green signal; 2.5 μM, 30 min), MitoTracker Green (MTG; green signal; 250 nM, 30 min), and MitoTracker Red CMX ROS (MTR; red signal; 250 nM, 30 min) fluorescent assays (Sigma-Aldrich), respectively. The analyses were performed using the protocol described in Gismondi et al. [49] (FACSCalibur instrument; filters: FL-1$^+$ for green; FL-2$^+$ for red) and counting 10,000 cytofluorimentric events per sample. Negative controls were carried out treating cells with PBS 1X, whereas positive controls were produced incubating cells with hydrogen peroxide (H$_2$O$_2$; 5 mM; for DCFH-DA test), S-nitrosoglutathione (GSNO; 0.5 mM: for DAF-FM DA test), and carbonyl cyanide m-chlorophenyl hydrazone (CCCP; 10 μM; for mitochondrial tests) for 4 h before the exposure to the appropriate probe. All results were reported as a percentage variation of cell fluorescence compared to the CNT sample, which was considered as a unit (100%). Changes in the mitochondrial membrane potential were reported as MTR/MTG ratio, as suggested by Pendergrass et al. [50].

### 2.10. Immunofluorescence Microscopy

For γH2AX and 53BP1 foci detection, cells (grown on slides) were fixed in 4% paraformaldehyde for 15 min, permeabilized with 0.4% Triton X-100 in PBS for 10 min, blocked in PBS blocking solution (10% FBS, 0.1% Triton X-100) for 3 h, and incubated for 2 h with primary antibodies (mouse monoclonal γH2AX Ser-139; rabbit polyclonal 53BP1; Merck Millipore). Then, samples were exposed for 1 h to the respective secondary antibodies (goat anti-mouse IgG labeled with Alexa Fluor 488 and goat anti-rabbit IgG labeled with Alexa Fluor 594; Invitrogen, Eugene, OR, USA). Nuclei were stained with 0.1 mg/mL of DAPI for 1 min. Images were acquired by a Leica DMR microscope (Leica Microsystems, Wetzlar, Germany) equipped with a Leica DFC 350 FX digital camera, EBQ 100 isolated fluorescent lamp (Leistungselektronik Jena GmbH, Jena, Germany), UV/FITC/TRITC filters, and 40X and 100X objectives. All images were elaborated by Leica Qwin Pro image analysis software and captured at the same instrument settings and exposure times in order to ensure a correct comparison. For foci counting, 500 cells for each experimental condition were analyzed by ImageJ. Control treatments were performed by etoposide (ETO; 500 nM, 8 h) and triethylenetetramine (TETA; 50 μM, 48 h) (Sigma-Aldrich).

### 2.11. Statistical Analysis

Results were reported as mean value ± standard deviation (SD) of measurements obtained by independent experiments ($n \geq 3$). Statistical significance was evaluated by one-way ANOVA test (Microsoft Excel software) vs. the respective control; a $p$-value $< 0.05$ was considered significant (* $< 0.05$; ** $< 0.01$; *** $< 0.001$).

## 3. Results

### 3.1. Chemical Characterization of the O. vulgare L. Phytocomplex

For a preliminary typization of the plant sample, the concentration of simple phenols and flavonoids in the *O. vulgare* L. extract was measured by spectrophotometric analyses. The amount of total phenols in HCOE was equal to 107.50 ± 10.81 μg GAE/g DMW, while flavonoids were 230.79 ± 13.97 μg QE/g DMW. Then, in order to identify the main plant metabolites underlying the bioactivity of the oregano extract, two different chromatographic approaches were applied to characterize the biochemical profile of this natural matrix. As reported in Table 1, 13 compounds were identified and quantified in HCOE by the HPLC-DAD technique (Supplementary Materials Figure S1).

**Table 1.** High-performance liquid chromatography-diode array detector (HPLC-DAD) profiles of oregano extract. Plant molecules and their concentration, detected in hydroalcoholic oregano extract (HCOE) by HPLC-DAD analysis, are reported. Results were indicated as μg of metabolite per 100 mg of dried material (μg/100 mg DMW) and represent the mean ± SD of six independent experiments.

| Compound | μg/100 mg DMW ± SD |
|---|---|
| Chrysin | 8.47 ± 0.06 |
| Rutin | 2.15 ± 0.05 |
| Myricetin | 0.03 ± 0.01 |
| Caffeic acid | 0.35 ± 0.02 |
| 1,1-Dimethylallyl caffeate | 1.28 ± 0.03 |
| Caffeic acid phenethyl ester | 0.76 ± 0.03 |
| Gallic acid | 0.18 ± 0.01 |
| Kaempferol | 0.04 ± 0.01 |
| *p*-Coumaric acid | 0.30 ± 0.01 |
| Genistein | 1.02 ± 0.02 |
| Quercetin-3-*o*-arabinoside | 2.37 ± 0.04 |
| Chlorogenic acid | 1.03 ± 0.04 |
| Apigenin | 0.26 ± 0.01 |

The most abundant molecule was chrysin (8.47 ± 0.06 µg/100 mg DMW), followed by quercetin-3-*o*-arabinoside (2.37 ± 0.04 µg/100 mg DMW) and rutin (2.15 ± 0.05 µg/100 mg DMW). Moreover, the phytocomplex extracted by *O. vulgare* L. samples was characterized by GC-MS analysis; in total, 45 secondary metabolites were detected and subjected to relative quantitation (Table 2).

**Table 2.** GC-MS profile of oregano extract. Plant metabolites and their relative abundance, detected in HCOE by GC-MS analysis, are reported. The relative abundance of each molecule was indicated as a percentage value with respect to the total mixture (100%). Values represented the mean of three independent experiments. The SD for each measurement was always <5% of the respective molecule peak area.

| GC-MS Detected Compound | % |
|---|---|
| *m*-Cymol | 0.57 |
| *p*-Mentha-1,3,8-triene | 0.31 |
| *p*-Cymene-2,5-dione | 0.87 |
| Thymol | 16.64 |
| Carvacrol | 34.82 |
| Caryophyllene oxide | 0.90 |
| *t*-Butylhydroquinone | 1.86 |
| Isopropyl laurate | 0.94 |
| Palmitic acid | 2.50 |
| Ethyl palmitate | 1.33 |
| Phytol | 0.84 |
| Retinoic acid | 0.73 |
| Methyl linolenate | 7.96 |
| Ethyl linolenate | 7.35 |
| 3,3-Dimethylbutanoic acid | 0.03 |
| *p*-Mentha-1,4-diene | 0.13 |
| *alpha*-Aminoisobutanoic acid | 0.03 |
| *p*-Mentha-6,8-dien-2-ol | 0.94 |
| Myrtenyl acetate | 0.54 |
| 2,6-Dimethyl-1,3,5,7-octatetraene | 0.29 |
| 3,5-Dihydroxy-6-methyl-2,3-dihydro-4H-pyran-4-one | 0.33 |
| Thujone | 0.52 |
| Octyl acetate | 0.19 |
| 2,3-Dimethyl-2-pentanol | 0.11 |
| Dimethyl malonate | 0.04 |
| Ehtyl acetimidate | 0.09 |
| Dimethylhexynediol | 0.20 |
| Linalool oxide | 0.89 |
| *trans*-2-Hexenyl caproate | 0.29 |
| 2-Ethyl-3-hydroxyhexyl 2-methylpropanoate | 0.09 |
| 1-Tetradecanol | 0.10 |
| Hydroxydehydrostevic acid | 0.11 |

**Table 2.** *Cont.*

| GC-MS Detected Compound | % |
| --- | --- |
| 2,2,4-Trimethyl-1,3-pentanediol diisobutyrate | 1.22 |
| Caprylic ether | 0.07 |
| *alpha*-Methylglucoside | 7.95 |
| Hexadecane | 0.92 |
| Retinyl acetate | 0.84 |
| Stearic acid | 0.99 |
| Butyl citrate | 3.58 |
| Jasmone | 0.44 |
| Oleic Acid | 0.33 |
| Oleic acid amide | 0.72 |
| 3-Hexadecanol | 0.40 |

The most abundant molecules were carvacrol (34.82%), thymol (16.61%), and linolenic acid methyl ester (7.96%).

### 3.2. O. vulgare L. Extract Reduces B16-F10 Cell Growth Not Affecting C2C12 Cell Viability

The biological effect of *O. vulgare* L. hydroalcoholic extract on the proliferation of B16-F10 cells, a murine melanoma line characterized by high aggressiveness and drug resistance, was investigated, by MTT assay, after exposure for 24 and 48 h with different concentrations of plant phytocomplex (0.1–10 mg/mL). Simultaneously, to check the safety of HCOE on non-tumor cells, C2C12 myoblasts were exposed to similar treatments. HCOE did not affect C2C12 cell growth after 24 h of incubation (Figure 1A), whereas a slight decrease of myoblast viability was observed after 48 h of treatment, reaching 27% at the highest concentration of extract (Figure 1B). By contrast, the oregano sample significantly decreased B16–F10 proliferation: in particular, after 24 and 48 h of incubation, 10 mg/mL of HCOE caused a reduction of melanoma cell viability of 73.42% (Figure 1C) and 84.11% (Figure 1D), respectively. According to these results, $IC_{50}$ values for C2C12 cells treated with HCOE, for 24 and 48 h, were estimated to be 55.44 and 14.28 mg/mL, respectively. For B16-F10, these values were 7.23 and 4.72 mg/mL, in that order. Consequently, for HCOE, the selectivity index was 7.66 after 24 h of treatment and 3.03 at 48 h.

The cytotoxicity of the oregano extract was evaluated by the Trypan Blue exclusion test. C2C12 and B16-F10 proliferation curves were generated, counting living cells after treatment with HCOE (2–10 mg/mL) or PBS (CNT) for 24 and 48 h. Simultaneously, dead cells were also counted in each sample. As expected, C2C12 cell growth was slightly affected by HCOE (Figure 1E). The strongest cytotoxic effect (21.25%) was observed after 48 h of incubation with the highest dose of HCOE. However, in all cases, the percentage of dead cells was always lower than 22% (Table 3). For B16-F10, cell proliferation significantly decreased in a dose-dependent manner (Figure 1F), and a remarkable percentage of dead cells was achieved at 10 mg/mL of HCOE (42% and 44.75% after 24 and 48 h of incubation, respectively) (Table 3).

To confirm the previous data, the B16–F10 cell cycle was analyzed after exposure to HCOE for 48 h. As shown in Figure 1G, low doses of plant extract induced a cell accumulation in the G1/G0 phase, whereas 10 mg/mL of HCOE determined an accumulation of cells in the G2/M phase equal to 45.24%.

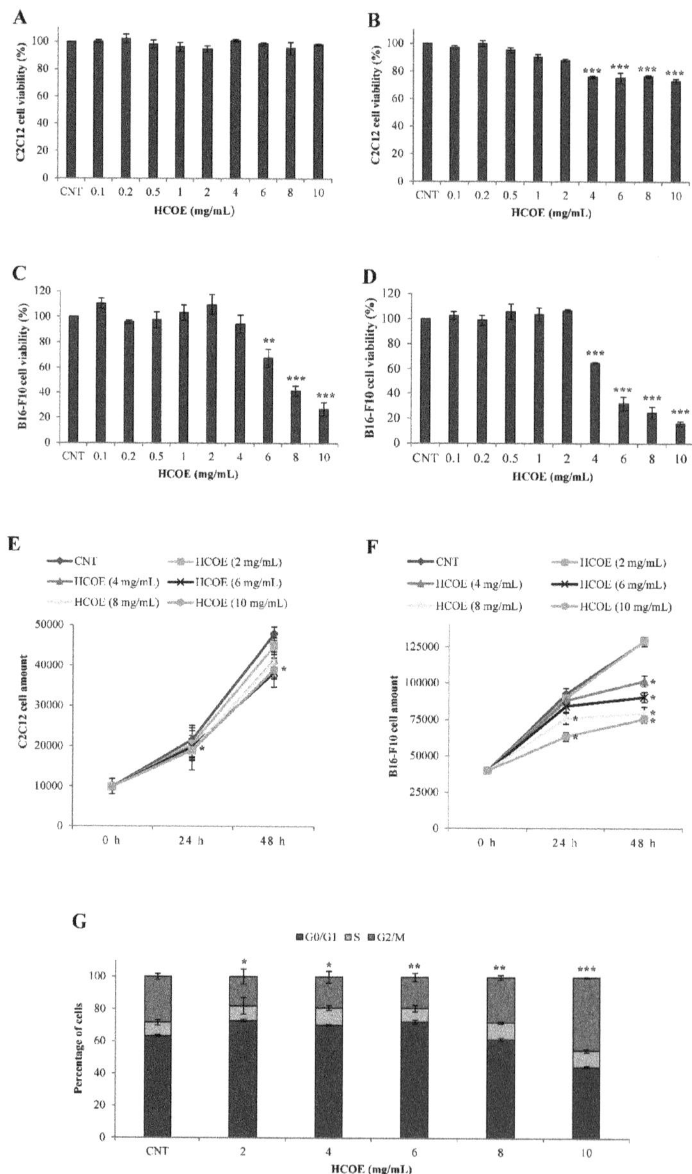

**Figure 1.** C2C12 and B16-F10 cell growth. C2C12 (**A**,**B**) and B16-F10 (**C**,**D**) cell growth was measured after 24 h (**A**,**C**) and 48 h (**B**,**D**) of treatments with Phosphate Buffered Saline (PBS), as control, and different concentrations of *O. vulgare* plant extract (HCOE). Results expressed as percentage with respect to PBS represent the mean ± SD of four independent experiments (* $p < 0.01$; *** $p < 0.001$ vs. control). Proliferation curves of C2C12 (**E**) and B16-F10 (**F**) cells were generated counting, by a Neubauer modified chamber, the amount of alive cells after staining with Trypan Blue at 0, 24, and 48 h of treatments with HCOE. Results are indicated as the mean ± SD of four independent experiments. (* $p < 0.05$ vs. control) (**G**) Cell cycle analysis of B16-F10 after treatment, for 48 h, with 2, 4, 6, 8, and 10 mg/mL HCOE is shown. For each sample, the percentage amount of cells in every cycle phase (G0/G1, S, and G2/M) was measured by cytofluorimetric analysis. Results are indicated as mean ± SD of four independent experiments (* $p < 0.05$; ** $p < 0.01$; *** $p < 0.001$ vs. control).

**Table 3.** Cytotoxicity analysis. The percentages of C2C12, B16-F10, and A375 dead cells identified by the Trypan Blue test after 24 and 48 h of HCOE treatment are reported. Data are reported as mean ± SD of four independent replicates. (* $p < 0.05$ vs. control).

| Cell Line | Treatment | | Time | |
|---|---|---|---|---|
| | | | 24 h | 48 h |
| | | CNT (PBS) | 6.07 ± 0.80 | 4.61 ± 0.11 |
| | | (2 mg/mL) | 13.33 ± 0.41 | 11.67 ± 2.60 |
| | | (4 mg/mL) | 13.16 ± 0.32 | 11.33 ± 3.43 |
| C2C12 | HCOE | (6 mg/mL) | 18.86 ± 0.28 | 16.35 ± 0.80 |
| | | (8 mg/mL) | 20.25 ± 0.51 * | 19.78 ± 1.57 * |
| | | (10 mg/mL) | 20.67 ± 0.17 * | 21.25 ± 2.36 * |
| | | CNT (PBS) | 1.86 ± 0.13 | 5.31 ± 0.26 |
| | | (2 mg/mL) | 14.90 ± 1.16 | 14.84 ± 1.17 |
| | | (4 mg/mL) | 12.98 ± 2.38 | 26.62 ± 0.95 * |
| B16-F10 | HCOE | (6 mg/mL) | 24.36 ± 0.84 * | 33.29 ± 1.49 * |
| | | (8 mg/mL) | 37.30 ± 2.90 * | 39.45 ± 3.44 * |
| | | (10 mg/mL) | 42.00 ± 3.73 * | 44.75 ± 1.86 * |
| | | CNT (PBS) | 5.38 ± 1.09 | 5.00 ± 1.71 |
| | | (2 mg/mL) | 17.78 ± 1.38 | 13.10 ± 1.92 |
| | | (4 mg/mL) | 16.82 ± 0.75 | 25.38 ± 2.96 * |
| A375 | HCOE | (6 mg/mL) | 29.44 ± 1.17 * | 34.71 ± 1.86 * |
| | | (8 mg/mL) | 31.21 ± 3.81 * | 48.26 ± 0.77 * |
| | | (10 mg/mL) | 47.14 ± 1.73 * | 55.13 ± 1.47 * |

### 3.3. Oregano Treatment Shows Antiproliferative Activity Also on A375 Human Melanoma Cells

The antiproliferative properties of the oregano extract on B16–F10 murine melanoma cells encouraged us to continue our research, testing if the same plant preparation could also exhibit similar effects on the A375 human melanoma line. The data obtained by the MTT assay are reported in Figure 2A,B. In detail, HCOE treatments at selected doses (2, 4, 6, 8, 10 mg/mL) decreased A375 cell viability, respectively, by 20.73%, 26.42%, 39.02%, 38.21%, 58.40%, and 3.86% after 24 h of incubation, and by 4.84%, 17.95%, 39.43%, 54.77%, and 80.40% after 48 h compared to the corresponding controls. In addition, in this case, the selectivity index was estimated. $IC_{50}$ values for A375 cells treated with HCOE, for 24 and 48 h, were 9.14 and 7.08 mg/mL, respectively. Consequently, the SI with respect to C2C12 cells was 6.07 after 24 h of treatment and 2.02 at 48 h.

The Trypan Blue exclusion test confirmed these results, evidencing reduced proliferation curves (Figure 2C) and significant cytotoxicity levels on human melanoma cells, especially after exposure to 10 mg/mL HCOE (47.14% and 55.13% of dead cells after 24 and 48 h of incubation, respectively) (Table 3).

Taking into account the great antiproliferative activity exerted by HCOE, the A375 cell cycle was analyzed after treatment with 10 mg/mL of oregano extract for 48 h. As reported in Figure 2D, a significant increase of cells in the G2/M phase (19.55%) was detected with respect to the control (CNT). In this context, the (TAX), a well-known plant drug able to induce G2/M phase arrest [51,52], was also used as positive control.

This evidence was consistent with the RT-PCR experiments performed to measure cyclin-dependent kinase 1 (CDK1), cyclin B1 (CCNB1), and P21 and P27 mRNA levels (Figure 2E). Indeed, the expression of CCNB1 and CDK1 genes, which are key factors in the transition from the G2 to the M phase [53], appeared reduced after 48 h of incubation with oregano extract. At the same time, P21 and P27 transcripts, which are CDK1/Cyclin B1 inhibitors [54–56], increased in the presence of HCOE. Similar results were obtained after exposure to TAX, although the P21 mRNA level remained unaltered compared to the control, as documented by the literature [57].

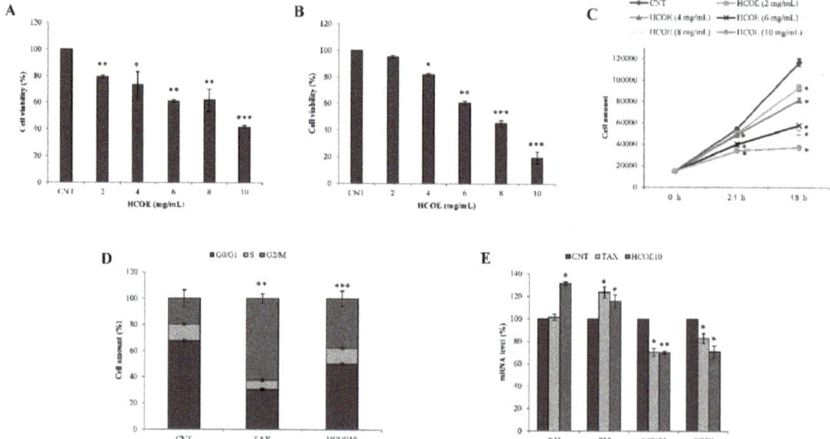

**Figure 2.** A375 cell growth and cycle. A375 cell viability was measured after 24 h (**A**) and 48 h (**B**) of treatment with 2, 4, 6, 8, and 10 mg/mL HCOE. Results, expressed as percentage with respect to the PBS control (CNT; considered as unit, 100%), represent the mean ± SD of four independent experiments (* $p < 0.05$; ** $p < 0.01$; *** $p < 0.001$ vs. control). (**C**) Proliferation curves of A375 cells were generated by counting, with a Neubauer modified chamber, the amount of alive cells after staining with Trypan Blue at 0, 24 and 48 h of treatment with HCOE. Results were indicated as mean ± SD of three independent experiments (* $p < 0.05$ vs. control). (**D**) A375 cell cycle analysis after 48 h of treatment with paclitaxel (TAX) or 10 mg/mL HCOE is shown. For each sample, the percentage amount of cells in every cycle phase (G0/G1, S and G2/M) was measured by cytofluorimetric analysis. Results are expressed as mean ± SD of three independent experiments (** $p < 0.01$; *** $p < 0.001$ vs. control). (**E**) P21, P27, CCNB1, and CDK1 mRNA levels, measured by RT-PCR, in A375 cells treated for 48 h with TAX or 10 mg/mL HCOE are reported. Gene expression, calculated as mRNA amount after normalization for ACTB mRNA ($2^{-\Delta\Delta Ct}$), is reported as percentage with respect to the PBS control (CNT; considered as unit, 100%). Data represent the mean ± SD of three independent measurements (* $p < 0.05$; ** $p < 0.01$ vs. control).

### 3.4. Oregano Extract Has Neither Mutagenic Nor Mutagen-Protective Effects

Before investigating, in depth, the antitumor effect exerted by HCOE on A375 cells, the potential mutagenic and mutagen-protective activities of the oregano extract were analyzed in order to confirm HCOE safety on non-tumor living model systems. For the assessment of both properties, an Ames test was carried out with *Salmonella typhimurium* strains TA97a, TA98, TA100, and TA1535 in the presence or absence of the metabolic activator S9 mix. As indicated in Supplementary Materials Table S2, HCOE did not show any toxicological evidence. Indeed, at all tested doses, the t/c values, namely the ratio between the number of colonies of *Salmonella* strains grown in the presence of oregano extract (t) and those on the control medium (c), were never higher or equal to 2 and never presented a dose–response trend [58].

As concerns the evaluation of the mutagen-protection effect mediated by HCOE, a properly modified Ames test [40] was carried out. The plant extract did not exhibit any protective activity against well-known mutagen compounds, as indicated by the inhibition rate (IR) percentages reported in Supplementary Materials Table S3. Indeed, despite the presence of IR-positive values, the HCOE dose–response effect was not observed with significant values.

### 3.5. Oregano Phytocomplex Impairs MITF Pathway and Accumulates Reactive Species in A375 Cells

The molecular mechanism underlying oregano antineoplastic activity was clarified by using A375 cells as a model system. In particular, according to previous data, 10 mg/mL of HCOE (thenceforth

HCOE10) was selected as a treatment dose, showing strong antiproliferative results on tumor cells and minimal effects on non-tumor ones.

The expression of the main genes involved in the MITF pathway, which is a cell signal that plays a crucial role in melanoma progression [59], was analyzed by RT-PCR (Figure 3A). In A375 cells, HCOE10 treatment, for 48 h, drastically decreased MITF, TYR, and TYRP1 mRNA levels of 16.27%, 27.93%, and 27.74%, respectively, compared to the control. This result was also corroborated by Western blotting analysis of Mitf protein content (Figure 3B); densitometric quantitation evidenced that oregano extract determined a reduction of 38.3% of Mitf protein, with respect to the control (Figure 3C).

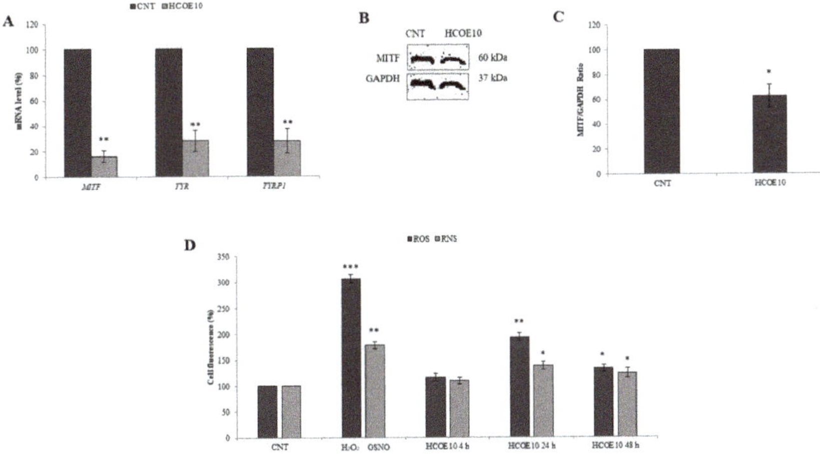

**Figure 3.** Analysis of mouse monoclonal microphthalmia-associated transcription factor (MITF) signaling and cell redox state. (**A**) *MITF*, tyrosinase (*TYR*), and tyrosinase-related protein 1 (*TYRP1*) mRNA levels, measured by RT-PCR, in A375 cells treated for 48 h with 10 mg/mL HCOE are reported. Gene expression calculated as mRNA amount after normalization for β-actin (ACTB) mRNA ($2^{-\Delta\Delta Ct}$) is reported as percentage with respect to the PBS control (considered as unit, 100%). Data represent the mean ± SD of four independent measurements (** $p < 0.01$ vs. control). (**B**) Representative Western blotting membrane of Mitf and Gapdh protein levels is shown. (**C**) Quantitation of MITF protein in A375 cells treated for 48 h with HCOE 10 mg/mL is reported. The results obtained by the ratio between Mitf and Gapdh (used as internal loading control) are indicated as percentage values with respect to PBS control (considered as unit, 100%). Data indicate the mean ± SD of three independent experiments (* $p < 0.05$ vs. control). (**D**) Intracellular reactive oxygen (ROS) and nitrogen (RNS) species levels were quantified in A375 cells, treated with 10 mg/mL of HCOE for 4, 24, and 48 h, by DCFH-DA and DAF-FM DA fluorescent assays, respectively. Radical species concentration is reported as percentage compared to PBS control (CNT). Hydrogen peroxide ($H_2O_2$) and S-Nitrosoglutathione (GSNO) treatments were performed as positive controls for ROS and RNS analysis, respectively. Results are expressed as mean ± SD of three independent measurements (* $p < 0.05$; ** $p < 0.01$; *** $p < 0.001$ vs. control).

As the inhibition of the MITF pathway has been associated to reactive species burst [29], the influence of the plant extract (that is HCOE10 treatment for 4, 24 and 48 h) on ROS and RNS levels was monitored in A375 by DCFH-DA and DAF-FM DA assays, respectively (Figure 4D). Oregano treatment for 4 h did not influence cell redox state, whereas a prolonged exposure caused a significant increase of reactive species: after 24 and 48 h, respectively, +92.82% and +32.51% for ROS and +37.42% and +22.98% for RNS, compared to control cells. Positive controls, using inducers of ROS (i.e., $H_2O_2$) and RNS (i.e., GSNO), were carried.

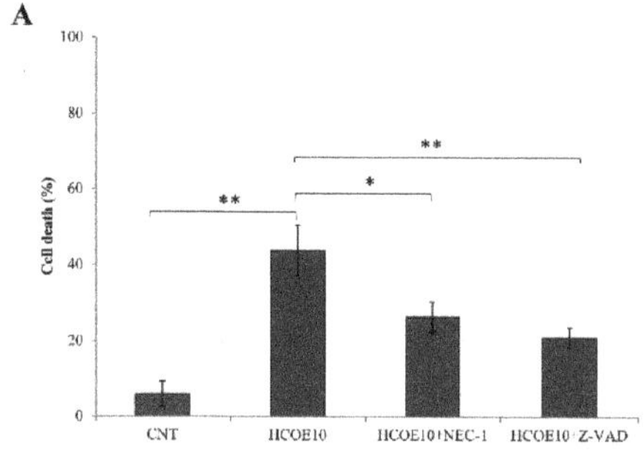

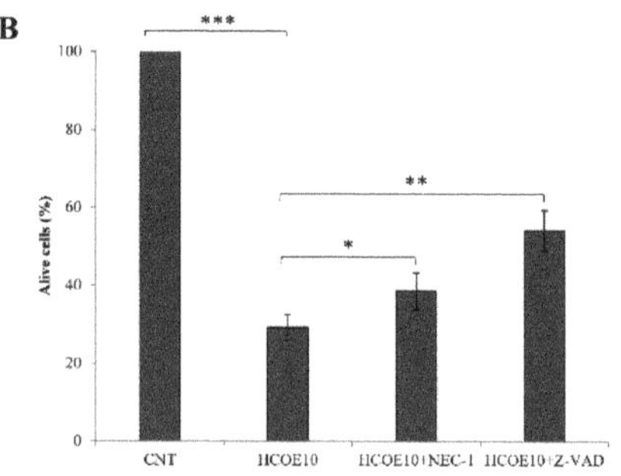

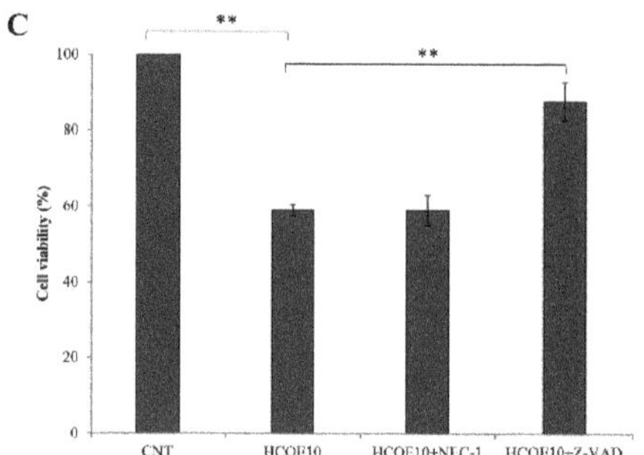

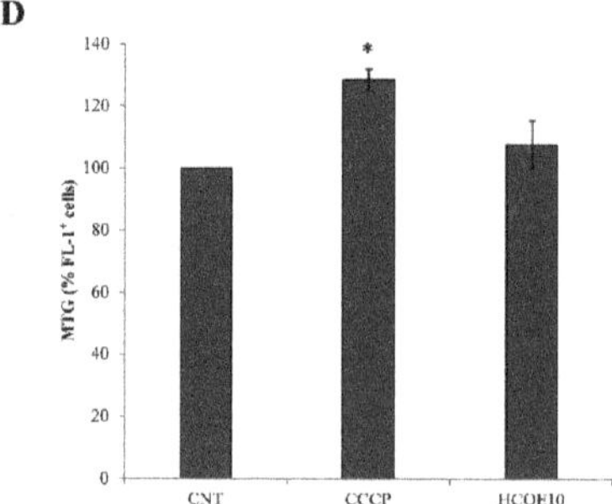

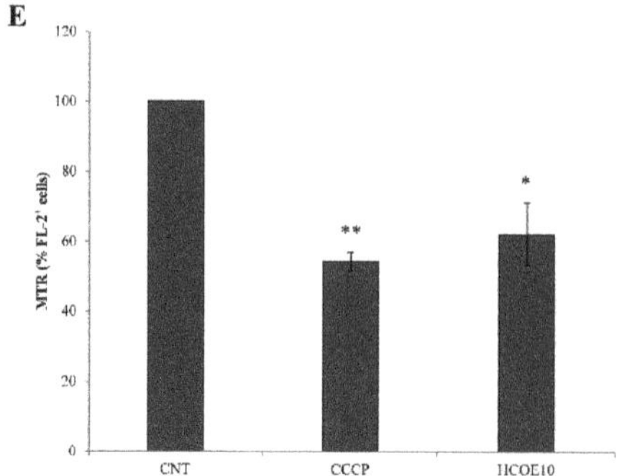

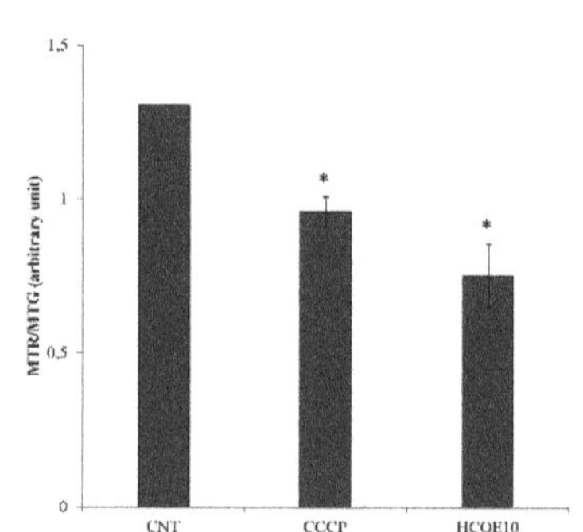

**Figure 4.** Cell death analysis and mitochondrial damage evaluation. A375 cell proliferation was analyzed after treatment, for 48 h, with 10 mg/mL HCOE, 10 mg/mL HCOE + 20 μM necrostatin-1 (HCOE10 + NEC-1), or 10 mg/mL HCOE + 20 μM Z-VAD-FMK (HCOE10+Z-VAD). (**A**) Cell death percentage was evaluated by flow cytometric assay, counting sub-G1 events. Results are expressed as mean ± SD of three independent experiments (* $p < 0.05$; ** $p < 0.01$ vs. control). (**B**) Alive cells were counted by the Trypan Blue exclusion test at 48 h of treatment with HCOE10, HCOE10+NEC-1, and HCOE10+Z-VAD. Results expressed as percentage with respect to the PBS control (CNT; considered as unit, 100%) are expressed as the mean ± SD of three independent experiments (* $p < 0.05$; ** $p < 0.01$; *** $p < 0.001$ vs. control). (**C**) A375 cell viability was measured by MTT assay after 48 h of treatment with HCOE10, HCOE10+NEC-1, and HCOE10+Z-VAD. Results, expressed as percentage with respect

to the PBS control (CNT; considered as unit, 100%), represent the mean ± SD of three independent experiments (** $p < 0.01$ vs. control). Mitochondrial mass (**D**) and membrane potential (**E**) measurements, after 4 h of treatment with carbonyl cyanide m-chlorophenyl hydrazone (CCCP) (10 μM) (used as positive control) or 48 h of incubation with 10 mg/mL HCOE, were carried out by using MitoTracker Green (MTG) and MitoTracker Red CMX ROS (MTR), respectively. Results are expressed as percentage variation of fluorescence with respect to PBS control (CNT; considered as unit, 100%) (* $p < 0.05$; ** $p < 0.01$ vs. control). (**F**) Changes in mitochondrial membrane potential are expressed as MTG/MTR ratio (* $p < 0.05$ vs. control).

### 3.6. Apoptosis/Necroptosis via Mitochondrial Pathway Is Induced in A375 Cells by HCOE

The decrease of melanoma cell growth, together with the evidence of high toxicity, cell cycle arrest, and MITF pathway inhibition, suggested that HCOE could induce cell death. For this reason, A375 cell viability was evaluated after co-treatments with oregano extract and Z-VAD-FMK (an anti-apoptotic pan-caspase inhibitor; [60]) or necrostatin-1 (NEC-1) (an inhibitor of necrosis/necroptosis [61]). Flow cytometry analysis (Figure 4A) showed that both co-treatments partially suppressed cell death (−17.43% of cells in the sub-G1 phase for the HCOE10 + NEC-1 sample; −22.90% for the HCOE10 + Z-VAD sample), with respect to the HCOE10-treated sample. Trypan Blue exclusion test (Figure 4B) confirmed this result, evidencing an increase of alive cells of 9.31% and 24.82% after exposure to HCOE10 + NEC-1 and HCOE10 + Z-VAD, respectively, compared to the treatment with only HCOE10. On the contrary, by MTT assay (Figure 4C), only HCOE10 + Z-VAD treatment seemed to rescue A375 cell viability compared to the HCOE10 sample.

Taking into account that mitochondrial damage is one of the main consequences of the oxidative stress (previously documented by ROS and RNS monitoring) and that the MTT assay (reported above) is an indicator of mitochondria activity [62,63], mitochondrial mass and membrane potential were estimated, respectively, by MitoTracker Green (MTG) and MitoTracker Red CMX ROS (MTR) cytofluorimetric assays in A375 exposed to HCOE10 for 48 h. CCCP, a well-known mitochondrial uncoupler, was used as a positive control. As shown in Figure 4D–F, HCOE10 treatment slightly affected mitochondrial mass, while it induced a strong depolarization. MitoTracker Red fluorescence was also normalized with the MitoTracker Green signal (MTR/MTG); this ratio was reduced by 42.17% after oregano treatment compared to the control.

Mitochondrial membrane permeability loss suggested mitochondrial damage and apoptosis induction. To check this hypothesis, Western blotting analyses were carried out, monitoring specific markers of these phenomena. In particular, the protein levels of Bax, Bcl-2, cytochrome c (Cycs), and mitochondrial import receptor subunit TOMM20 were detected and quantified (Figure 5A–C). After 48 h of exposure to HCOE10, an approximately 2-fold increase of Bax/Bcl-2 ratio was observed. Moreover, as demonstrated in the case of staurosporin treatment (STS, used as positive control) [64], the plant extract caused the increase of Cycs (+135.2%) compared to TOMM20. The lack of changes in the TOMM20 level in oregano-treated cells confirmed the previous cytofluorimetric results for mitochondria mass..

Finally, additional immunoblots (and relative densitometric quantitations) were performed to study caspase-3 (Casp-3) and Parp-1 levels, as shown in Figure 5D–F. HCOE10 treatment, for 48 h, significantly decreased pro Casp-3 and full-length Parp-1 levels with respect to the negative control (CNT), inducing Parp-1 cleavage as also observed in the presence of STS. These effects were almost completely rescued by treating melanoma cells with HCOE10 and Z-VAD simultaneously.

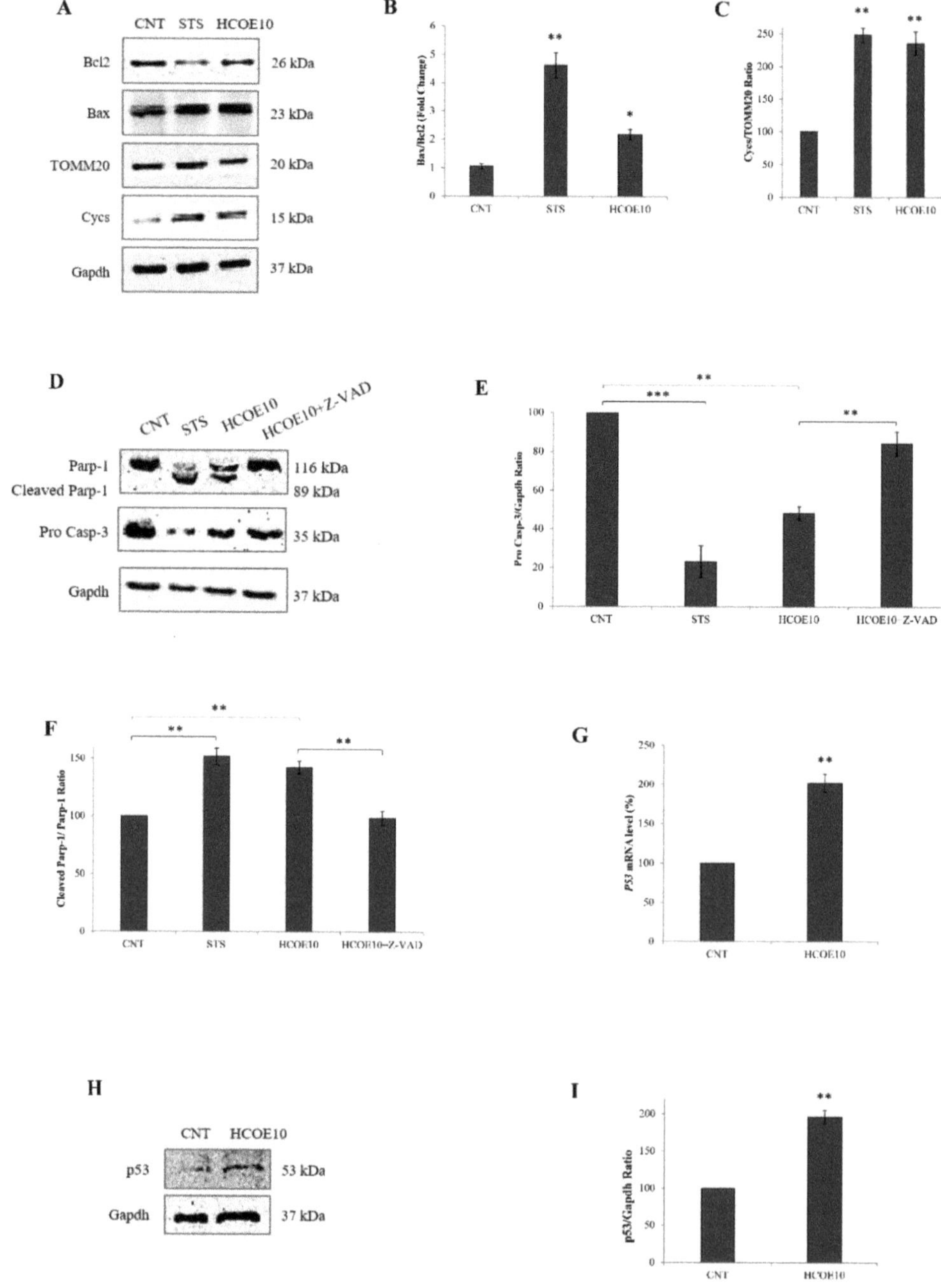

**Figure 5.** Analysis of mitochondria, apoptosis, and DNA damage markers. (**A**) Representative Western blotting membranes of Bcl-2, Bax, TOMM20, Cycs, and Gapdh protein levels evaluated in A375 cells treated for 4 h with staurosporine (STS, 2 µM) (used as positive control) or for 48 h with 10 mg/mL HCOE are shown. Quantitation of Bcl-2, Bax, TOMM20, and Cycs proteins was performed and the ratio

between Bax/Bcl-2 (**B**) and Cycs/TOMM20 (**C**) are reported as percentage values, with respect to PBS control (CNT; considered as unit, 100%). (**D**) Representative Western blotting membranes of full length and cleaved form of Parp-1, pro caspase-3 (pro Casp-3), and Gapdh protein levels are shown. Quantitation of pro Casp-3 and Parp-1 proteins in A375 cells treated for 4 h with STS or for 48 h with 10 mg/mL HCOE and 10 mg/mL HCOE + 20 $\mu$M Z-VAD-FMK (HCOE10 + Z-VAD) are reported. (**E**) Pro Casp-3 levels evaluated by the ratio between pro Casp-3 and Gapdh (used as internal loading control) are expressed as percentage values with respect to PBS control (CNT; considered as unit, 100%). (**F**) The cleaved Parp-1/full-length Parp-1 ratio is expressed as percentage values with respect to PBS control (CNT; considered as unit, 100%). (**G**) P53 mRNA level were measured by RT-PCR in A375 cells treated for 48 h with 10 mg/mL of HCOE. Gene expression, calculated as transcript amount after normalization for ACTB mRNA ($2^{-\Delta\Delta Ct}$), is reported as percentage with respect to the PBS control (CNT; considered as unit, 100%). (**H**) A representative Western blotting membrane of p53 and Gapdh protein levels is shown. (**I**) Quantitation of p53 protein in A375 cells treated for 48 h with 10 mg/mL HCOE is reported. Results obtained by the ratio between p53 and Gapdh (used as loading control) signals are indicated as percentage values with respect to PBS control (CNT; considered as unit, 100%). All data indicate the mean $\pm$ SD of three independent experiments (* $p < 0.05$; ** $p < 0.01$; *** $p < 0.001$ vs. negative control).

### 3.7. HCOE Triggers DNA Breakages Mediated by Metal Ions

Since the pro-oxidant activity of several plant metabolites has been associated to DNA damage [65], the P53 gene expression level was monitored both in terms of transcript and protein amount, in A375 cells after exposure to HCOE10, for 48 h. As shown in Figure 5G–I, oregano treatment increased p53 mRNA (+102.1%) and protein (+95.95%) concentration, compared to the control.

To confirm the induction of DNA breakages by oregano extract, $\gamma$H2AX and 53BP1 foci, two well-known markers of DNA damage [66,67], were detected by immunofluorescence (IF) analysis on A375 cells treated with HCOE10 for 48 h. In detail, an average of 14.86 $\pm$ 1.50 $\gamma$H2AX foci and 14.10 $\pm$ 1.05 53BP1 foci per cell were measured with respect to 2.60 $\pm$ 1.56 $\gamma$H2AX foci and 3.27 $\pm$ 1.14 53BP1 foci per cell found in the control sample (Figure 6A). Representative IF images per each sample were reported in Figure 6B; here, the treatment with etoposide (ETO), an inhibitor of topoisomerase II enzyme, represented the positive control. IF analysis showed that the major part (82% of cases) of $\gamma$H2AX and 53BP1 foci co-localized (Figure 6B, panel l). In addition, considering that polyphenols, such as flavonoids, catalyze DNA breakages in the presence of metal ions (e.g., copper) [68], we evaluated the ability of oregano extract to trigger DNA damage in the presence of a copper chelator (TETA). The co-treatment HCOE10 + TETA determined a significant reduction of the level of DNA breakages (9.91 $\pm$ 2.91 $\gamma$H2AX foci and 5.71 $\pm$ 1.87 53BP1 foci per cell) with respect to the pure treatment with HCOE10 (Figure 6A,B, panels q–t).

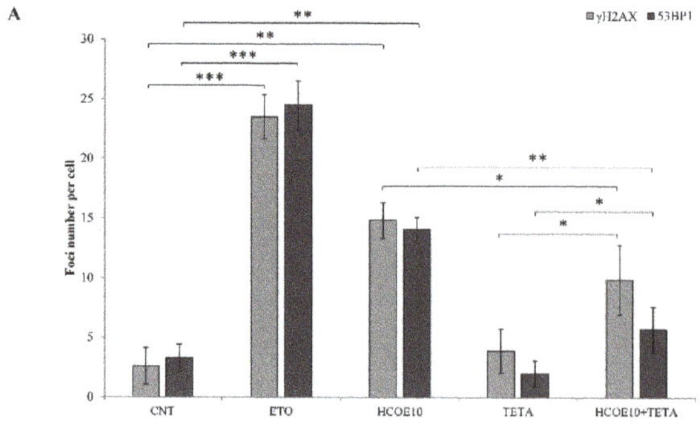

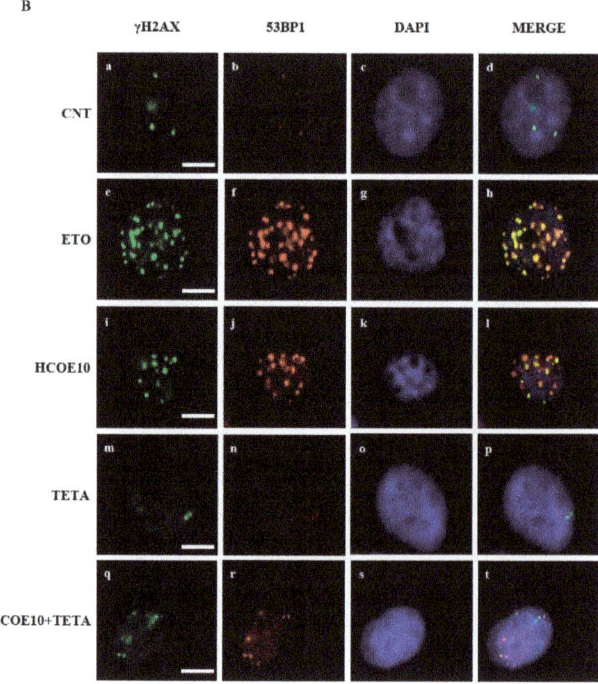

**Figure 6.** Immunofluorescence analysis. (**A**) The amount of γH2AX and 53BP1 foci detected in A375 cells treated with PBS (CNT), etoposide (ETO), 10 mg/mL HCOE (HCOE10), triethylenetetramine (TETA) or HCOE10+TETA were counted. Results are expressed as mean of foci per cell ± SD of three independent experiments (* $p < 0.05$; ** $p < 0.01$; *** $p < 0.001$ vs. control). (**B**) Representative immunofluorescence images of A375 cells treated with PBS (CNT) (**a–d**), etoposide (ETO) (**e–h**), 10 mg/mL HCOE (HCOE10) (**i–l**), triethylenetetramine (TETA) (**m–p**), or HCOE10+TETA (**q–t**) are shown. γH2AX foci are in green, 53BP1 foci are in red, while nuclei were stained in blue with DAPI. Merged images are also reported. The white bars indicate 15 μm.

## 4. Discussion

Among tumors, skin cancer is the most common neoplasia worldwide. In particular, the more aggressive and deadliest form of this pathology is represented by melanoma [69]. Melanoma is a multi-factorial disease, depending on both environmental and endogenous factors. Indeed, about 90% of melanomas are caused by ultraviolet light exposure [70], while the remaining 10% has been associated to genetic defects [71]. Currently, such type of skin cancer is treated by surgical removal, which leads to a high survival rate except in the presence of metastases. In the latter case, a chemotherapeutic approach based on several drugs, such as Dacarbazine (an alkylating agent), Vemurafenib (BRAF kinase inhibitor), Ipilimumab (monoclonal antibody targeting for cytotoxic T-lymphocyte antigen-4), Pemrolizumab (monoclonal IgG4 antibody), and Nivolumab (monoclonal antibody targeting for Programmed Death-1 protein) [72], is the most efficient strategy to treat melanoma. However, the aggressiveness and the high rate of multi-drug resistance of this pathology highlight the need of new antineoplastic molecules.

Phytochemicals, namely secondary metabolites, produced by plants to protect themselves from environmental stresses and promote their reproduction [73], have been widely documented to exert a great variety of non-negligible bioactivities even on mammalian systems. Indeed, it has been documented that several plant compounds promote apoptosis and inhibit metastasis and angiogenesis [74–77]. For this reason, they have been taken under consideration thanks to their relevant medical and pharmaceutical properties. In this scenario, plant phytocomplexes can represent potential antiproliferative and anti-invasive cocktails for drug-resistant melanomas.

Based on the previous evidence, in the current research, the biological effect of a hydroalcoholic extract from plants of *Origanum vulgare* L. ssp. *hirtum* (HCOE) was investigated on highly metastatic and drug-resistant murine (B16-F10) and human (A375) melanoma cells. Indeed, although a potential antineoplastic effect has been associated to oregano extracts [30–35], the capacity of this herb to contrast the growth of the above-mentioned melanoma lines has never been elucidated in detail. Moreover, as one of the main goals of the cancer research is the discovery of new drugs with limited or without adverse side effects for healthy tissues, oregano extract was also tested on C2C12 myoblasts, which is a non-tumor cell model.

Since plant extract bioactivity cannot be attributed only to its more representative compounds, rather than to the synergic effect of plant molecules present both in high concentration and in trace [78–80], first of all, we investigated the biochemical profile of the oregano extract used in this study by chromatographic approaches (HPLC-DAD and GC-MS). A total of 58 metabolites were detected and recognized. Moreover, the amount of total phenols and flavonoids in HCOE was also measured, together with its in vitro antiradical power, to better characterize the plant extract.

The GC-MS chemoprofile obtained in the current research totally was in line with those documented in the literature. In this regard, *O. vulgare* ssp. *hirtum* essential oil can be classified in four chemotypes, according to its main constituents, especially thymol and carvacrol, and relative ratios. For instance, *O. vulgare* ssp. *hirtum* essential oil extracted from plants grown in Southern Italy and Northern Greece would seem rich in thymol, while that obtained from plant material propagated in Southern Greece was abundant in carvacrol [81–83]. Although purified from plants grown on Mount Athos (Northern Greece), our oregano extract showed a chemoprofile more similar to those of Southern Greece. To explain this phenomenon, it is important to keep in mind that several independent variables, such as plant growth stage and environment conditions, may strongly influence the phytocomplex. On the other hand, the oregano sample here studied revealed a content of phenolics double compared to that reported in the literature [84,85].

As concerns oregano biological activity, the plant phytocomplex determined a great reduction of B16-F10 cell growth, especially after 48 h of exposure with the highest doses (6, 8, 10 mg/mL), while it minimally influenced the myoblast division rate. B16-F10 and C2C12 proliferation curves confirmed the previous MTT outcomes. Moreover, HCOE induced a significant time- and dose-dependent toxicity on murine melanoma cells, whereas a low percentage of Trypan blue positive cells was detected in the case of C2C12. To validate these results, $IC_{50}$ values and selectivity indexes (SI) for the plant treatments on both cell lines were calculated. According to the literature, a reliable SI value must be equal to or greater than 2 [86], and HCOE satisfied this requirement. Cell cycle analysis of B16-F10 demonstrated that oregano extract caused an increase of cells in G2/M phase.

This promising evidence encouraged us to check if oregano extract could exert antineoplastic effects also against a human melanoma cell line with the aim to lay the basis for future desirable application in translational medicine. As expected, significant antiproliferative activity, with a high selectivity index, together with a relevant cytotoxic effect of HCOE on A375 cells was confirmed.

Taking into account all previous data, the concentration of 10 mg/mL of oregano sample was selected for further experiments, showing the best antiproliferative effect.

Cytofluorimetric and RT-PCR analyses proved that the plant extract blocked cell division in the G2/M phase, acting on the expression of specific key genes implicated in the inception of the mitotic process (i.e., CDK1, CCNB1, P21, and P27) [53,55,56] such as paclitaxel, which is a well-known plant anticancer drug with antimitotic property.

Before proceeding with the other analyses, as the final objective of the current research was the valorization of an oregano hydroalcoholic extract for potential chemotherapeutic applications, the control of the safety for the plant preparation with reference to mutagenic properties was necessary. For this purpose, the Ames test, recommended by the European Food Safety Authority (EFSA) as a proper assay to assess food safety [87], was carried out. It provided the proof that HCOE did not have a non-mutagenic effect [88,89], as expected. Indeed, *O. vulgare* and its derivatives, which are accepted as food ingredients by the U.S. Food and Drug Administration, are listed among the GRAS (Generally Recognized As Safe) substances by the Code of Federal Regulations of the USA and are generally well tolerated by the human body, although gastrointestinal upset and skin allergic reactions have been associated to them [90]. However, the lack of data regarding oregano genotoxicity [91,92] highlights the need for further studies on this topic and valorizes the present preliminary results. The mutagen-protective activity of HCOE was also investigated in order to further analyze the plant extract under a healthy point of view. Unfortunately, at all tested doses, no protective property against known mutagenic compounds was documented.

In the second part of this work, the molecular mechanism underlying the bioactivity of HCOE on melanoma cells was investigated in depth. First of all, the efficiency of the MITF pathway was verified, studying the expression rate of MITF, TYR, and TYRP1 genes. The results revealed an antimelanogenic activity of the oregano extract on A375, as already suggested by the literature [93].

At low levels, reactive species play a key role in cell signaling, but their overproduction can lead to mitochondrial alterations (i.e., DNA mutations, respiratory chain damage, membrane permeability loss), oxidative stress, and an inhibition of specific genes, including those related to melanin synthesis [29,94]. Therefore, according to the previous results, intracellular ROS and RNS levels were measured in A375 cells, demonstrating that HCOE10 had a remarkable pro-oxidant effect, especially after 24 h of incubation. This evidence suggested that oregano antitumor activity could be based on a reactive species-mediated apoptotic process. Literature data about carvacrol bioactivity, one of the most abundant phenolic monoterpenoids of oregano (as also documented in the present research by GC-MS analysis, see Table 2), would support this hypothesis. Indeed, several published works showed that this compound induces apoptosis in different tumor cell lines by increasing the ROS amount and disrupting mitochondrial membrane potential [95–97].

All previous considerations suggested that HCOE10 could trigger cell death in A375. For this reason, to clarify which cell death pathway was induced by oregano extract, a pan-caspase inhibitor, Z-VAD-FMK, and an inhibitor of necroptosis, necrostatin-1 [98], were used in co-treatments with HCOE10 on the melanoma cells. The experiments demonstrated that the decrease of tumor cell proliferation was partially due to both apoptosis and necroptosis induction, as already suggested by Savini et al. [99] and Rubin et al. [100]. Indeed, these two phenomena are strongly correlated to each other because they share the same stimuli (such as TNF-α), ligands, and receptors [101,102]. The mechanism underlying the activation of apoptosis and/or necroptosis is still under investigation. Nevertheless, Annexin V/propidium iodide staining in the presence of necrostatin-1, as well as the analysis of TNF-α level, could be performed in the future to better clarify the role of these two pathways in oregano-induced cell death [103,104].

Since in MTT assay, only HCOE10 + Z-VAD double treatment rescued A375 viability, taking into account that mitochondria are both generators of and targets for reactive species (whose levels were previously observed to be increased in the presence of HCOE), a mitochondria impairment caused by oregano extract was hypothesized. To validate this theory, cytofluorimetric analyses, based on the use of MitoTracker Green and MitoTracker Red CMX-ROS probes, were performed. The data confirmed a loss of mitochondrial potential after HCOE10 treatment, while no significant change in mitochondrial mass was appreciated. By contrast, CCCP-treated cells (the positive control) showed an increase of mitochondrial mass, which was probably due to mitochondrial fragmentation [105,106] associated to an expected membrane depolarization.

As known in the literature, the permeabilization of the mitochondrial outer membrane is an event promoted by the pro-apoptotic protein Bax. Bcl-2, on the other hand, is an anti-apoptotic factor that prevents apoptosis by inhibiting Bax. Therefore, an elevated Bax/Bcl-2 ratio is a feature of apoptotis induction. As a consequence of Bax activation, damaged mitochondria release cytochrome c into the cytoplasm, leading to caspase-3 induction and Parp-1 cleavage [107,108]. In view of this, the amount of apoptosis-related proteins (Bax, Bcl-2, Cycs, Casp-3, and Parp-1) were evaluated in A375 after 48 h of incubation with HCOE10. Oregano extract was able to trigger apotosis in the human melanoma cells by increasing Bax and Cycs concentrations and decreasing Bcl-2 and pro Casp-3 amounts. Moreover, as expected, Parp-1 cleavage was evident in oregano-treated cells. HCOE10 + Z-VAD co-treatment reversed HCOE effect, showing protein levels similar to those detected in the respective negative control. All these results confirmed that the plant extract induced, in A375 cells, a caspase-dependent apoptosis, which was mediated by mitochondrial damage. However, considering that both ROS and RNS play an important role also in autophagy and that mitophagy is strictly involved in mitochondrial turnover [109,110], it would be interesting in the future to investigate if autophagy/mitophagy is also induced by oregano treatment.

Another important key element in the apoptotic process is the P53 gene, whose protein promotes BAX gene expression by direct activation of its promoter and BCL2 downregulation [111–113]. P53 is activated in response to a wide range of genotoxic insults. It is involved in several DNA-repair machineries, such as nucleotide excision repair, for the removal of helix-distorting lesions (typical of UV-damage) and base excision repair (BER) in case of base oxidative modifications [114,115]. As known in the literature, plant compounds, especially polyphenols, can cause oxidative DNA strand breakage, alone or in the presence of transition metal ions. Indeed, among all, the copper (that is the most abundant ion of the cell nucleus, together with zinc), after reduction from Cu(II) to Cu(I) form by the action of plant metabolites, is particularly prone to produce ROS (especially the hydroxyl radical) during its re-oxidation, bind chromatin (particularly guanines), and cause DNA breakages [116,117].

The existence of such types of event prompted us to analyze P53 mRNA and protein levels and evaluate DNA damages (detecting γH2AX and 53BP1 foci by IF) upon HCOE10 exposure. Oregano treatment induced in A375 p53 upregulation, at both the transcriptional and translational level, and DNA breakages. In detail, γH2AX is the phosphorylated form of the histone 2AX (H2AX). Its phosphorylation is an early consequence of double and single-strand breakages [118–121]; therefore, the detection of γH2AX is widely used as a marker of DNA damage. On the other hand, p53 binding protein 1 (53BP1) locates only to DNA double-strand breaks [122]. IF analyses highlighted a great number of γH2AX and 53BP1 foci, which often co-localized in HCOE10-treated cells with respect to untreated controls. This evidence, together with Ames test results, pointed out that oregano hydroalcoholic extract acts as a genotoxic but not mutagenic agent in A375 melanoma cells, causing DNA single and double-strand breaks [123]. Furthermore, taking into account that DNA damage often results from the binding of phytochemical to transition metal ions, such as copper [124–126], co-treatments with HCOE10 and a copper chelator (TETA) were performed. Surprisingly, the sequestration of copper by TETA protected A375 cells from HCOE10-induced DNA damage, confirming that this metal ion plays a fundamental role in oregano bioactivity and relative DNA cleavage reaction.

In cancer masses, the concentration of iron and zinc is lower, whereas the copper concentration is usually higher than in healthy tissues [127–132]. This feature can explain why plant compounds exert selective cytotoxic activity against tumor cells but not toward non-tumor ones [133–139], justifying our data about oregano cytotoxicity on B16–F10 and A375 but not on C2C12.

In conclusion, *O. vulgare* hydroalcoholic extract, due to its peculiar chemical profile and pro-oxidant effect, inhibits melanogenesis and melanoma cell proliferation. Through in-depth molecular analyses, the antineoplastic activity of the oregano extract was associated to its ability to trigger programmed cell death (apoptosis and necroptosis) in A375 human melanoma cells via mitochondria and DNA damage. As this molecular mechanism was correlated to the intracellular/nuclear concentration of copper ions, oregano phytochemicals appeared to be slightly toxic or non-toxic for non-tumor cells. All this evidence represents a robust starting point for further studied focused on the design of new anti-melanoma natural drugs. Indeed, according to the present data, *O. vulgare* phytocomplex, working in synergy, represent an excellent candidate as anticancer agent, being highly selective and effective against human melanoma cells.

**Supplementary Materials:** The following are available online at http://www.mdpi.com/2304-8158/9/10/1486/s1, Figure S1: HPLC analysis; Table S1: RT-qPCR primers; Table S2: Ames test for mutagenic analysis; Table S3: Ames test for mutagen-protective analysis.

**Author Contributions:** Conceptualization: V.N., A.G. and G.S.; Methodology: A.G.; Formal analysis: V.N., G.D.M. and G.S.; Writing—Original draft preparation: V.N. and A.G.; Writing—Review and editing: all authors; Resources: A.C., G.S. and A.G.; Supervision: Antonella Canini and A.G. All authors have read and agreed to the published version of the manuscript.

**Funding:** This research received no external funding.

**Acknowledgments:** The authors want to thank Marco Sarandrea (SARANDREA MARCO & CO. srl) who performed the sampling of the plant material and monks of MOUNT ATHOS VATOPEDI HOLY MONASTERY (Ιερά Μονή Βατοπεδίου) in Greece who kindly provided us oregano plants. Moreover, we also thank Alessandro Grandini for technical assistance performing the Ames test, Miss Sophie Gart for the revision of the English, and Maria Rosa Ciriolo, Katia Aquilano, Carlo Rodolfo and Daniele Lettieri Barbato for their suggestions on mitochondrial analyses. This research did not receive any specific grant from funding agencies in the public, commercial, or not-for-profit sectors.

**Conflicts of Interest:** The authors declare no conflict of interest.

# References

1. Clardy, J.; Walsh, C. Lessons from natural molecules. *Nature* **2004**, *432*, 829–837. [CrossRef] [PubMed]
2. Newman, D.J.; Cragg, G.M. Natural products as sources of new drugs from 1981 to 2014. *J. Nat. Prod.* **2016**, *79*, 629–661. [CrossRef] [PubMed]
3. Patridge, E.; Gareiss, P.; Kinch, M.S.; Hoyer, D. An analysis of FDA-approved drugs: Natural products and their derivatives. *Drug Discov. Today* **2016**, *21*, 204–207. [CrossRef] [PubMed]
4. Thomford, N.; Senthebane, D.; Rowe, A.; Munro, D.; Seele, P.; Maroyi, A.; Dzobo, K. Natural products for drug discovery in the 21st century: Innovations for novel drug discovery. *Int. J. Mol. Sci.* **2018**, *19*, 1578. [CrossRef]
5. Gordaliza, M. Natural products as leads to anticancer drugs. *Clin. Transl. Oncol.* **2007**, *9*, 767–776. [CrossRef] [PubMed]
6. Greenwell, M.; Rahman, P.K. Medicinal plants: Their use in anticancer treatment. *Int. J. Pharm. Sci. Res.* **2015**, *6*, 4103–4112. [PubMed]
7. Singh, S.; Sharma, B.; Kanwar, S.S.; Kumar, A. Lead phytochemicals for anticancer drug development. *Front. Plant Sci.* **2016**, *7*, 1667. [CrossRef]
8. Gullett, N.P.; Ruhul Amin, A.R.M.; Bayraktar, S.; Pezzuto, J.M.; Shin, D.M.; Khuri, F.R.; Aggarwal, B.B.; Surh, Y.J.; Kucuk, O. Cancer prevention with natural compounds. *Semin. Oncol.* **2010**, *37*, 258–281. [CrossRef] [PubMed]
9. Lesgards, J.F.; Baldovini, N.; Vidal, N.; Pietri, S. Anticancer activities of essential oils constituents and synergy with conventional therapies: A review. *Phytother. Res.* **2014**, *28*, 1423–1446. [CrossRef]

10. Bharti, A.C.; Aggarwal, B.B. Nuclear factor-kappa B and cancer: Its role in prevention and therapy. *Biochem. Pharmacol.* **2002**, *64*, 883–888. [CrossRef]

11. Newman, D.J.; Cragg, G.M.; Holbeck, S.; Sausville, E.A. Natural products as leads to cell cycle pathway targets in cancer chemotherapy. *Curr. Cancer Drug Targets* **2002**, *2*, 279–308. [CrossRef] [PubMed]

12. Surh, Y.J. Cancer chemoprevention with dietary phytochemicals. *Nat. Rev. Cancer* **2003**, *3*, 768–780. [CrossRef] [PubMed]

13. Hollósy, F.; Kéri, G. Plant-derived protein tyrosine kinase inhibitors as anticancer agents. *Curr. Med. Chem. Anticancer Agents* **2004**, *4*, 173–197. [CrossRef] [PubMed]

14. Paul, M.K.; Mukhopadhyay, A.K. Tyrosine kinase—Role and significance in Cancer. *Int. J. Med. Sci.* **2004**, *1*, 101–115. [CrossRef] [PubMed]

15. Cragg, G.M.; Newman, D.J. Plants as a source of anti-cancer agents. *J. Ethnopharmacol.* **2005**, *100*, 72–79. [CrossRef] [PubMed]

16. Nam, N.H. Naturally occurring NF-kappaB inhibitors. *Mini Rev. Med. Chem.* **2006**, *6*, 945–951. [CrossRef] [PubMed]

17. Huang, S. Inhibition of PI3K/Akt/mTOR signaling by natural products. *Anticancer Agents Med. Chem.* **2013**, *13*, 967–970. [CrossRef] [PubMed]

18. Weaver, B.A. How Taxol/paclitaxel kills cancer cells. *Mol. Biol. Cell* **2014**, *25*, 2677–2681. [CrossRef]

19. Kaur, S.; Kumar, M.; Kaur, S. Role of phytochemicals in MAPK signaling pathway-mediated apoptosis: A possible strategy in cancer chemoprevention. *Stud. Nat. Prod. Chem.* **2016**, *50*, 159–178.

20. Low, H.B.; Zhang, Y. Regulatory roles of MAPK phosphatases in cancer. *Immune Netw.* **2016**, *16*, 85–98. [CrossRef]

21. Nosrati, N.; Bakovic, M.; Paliyath, G. Molecular mechanisms and pathways as targets for cancer prevention and progression with dietary compounds. *Int. J. Mol. Sci.* **2017**, *18*, 2050. [CrossRef] [PubMed]

22. Suvarna, V.; Murahari, M.; Khan, T.; Chaubey, P.; Sangave, P. Phytochemicals and PI3K inhibitors in cancer-An insight. *Front. Pharmacol.* **2017**, *8*, 916. [CrossRef] [PubMed]

23. Valko, M.; Leibfritz, D.; Moncol, J.; Cronin, M.T.D.; Mazur, M.; Telser, J. Free radicals and antioxidants in normal physiological functions and human disease. *Int. J. Biochem. Cell Biol.* **2007**, *39*, 44–84. [CrossRef] [PubMed]

24. Hadi, S.M.; Asad, S.F.; Singh, S.; Ahmad, A. Putative mechanism for anticancer and apoptosis-inducing properties of plant-derived polyphenolic compounds. *IUBMB Life* **2000**, *50*, 167–171. [PubMed]

25. Elbling, L.; Weiss, R.M.; Teufelhofer, O.; Uhl, M.; Knasmueller, S.; Schulte-Hermann, R.; Berger, W.; Micksche, M. Green tea extract and (-)-epigallocatechin-3-gallate, the major tea catechin, exert oxidant but lack antioxidant activities. *FASEB J.* **2005**, *19*, 807–809. [CrossRef] [PubMed]

26. Aquilano, K.; Baldelli, S.; Rotilio, G.; Ciriolo, M.R. Role of nitric oxide synthases in parkinson's disease: A review on the antioxidant and anti-inflammatory activity of polyphenols. *Neurochem. Res.* **2008**, *33*, 2416–2426. [CrossRef]

27. Martin-Cordero, C.; Jose Leon-Gonzalez, A.; Manuel Calderon-Montano, J.; Burgos-Moron, E.; Lopez-Lazaro, M. Pro-oxidant natural products as anticancer agents. *Curr. Drug Targets* **2012**, *13*, 1006–1028. [CrossRef] [PubMed]

28. Khan, H.Y.; Zubair, H.; Faisal, M.; Ullah, M.F.; Farhan, M.; Sarkar, F.H.; Ahmad, A.; Hadi, S.M. Plant polyphenol induced cell death in human cancer cells involves mobilization of intracellular copper ions and reactive oxygen species generation: A mechanism for cancer chemopreventive action. *Mol. Nutr. Food Res.* **2014**, *58*, 437–446. [CrossRef]

29. Nanni, V.; Canuti, L.; Gismondi, A.; Canini, A. Hydroalcoholic extract of *Spartium junceum* L. flowers inhibits growth and melanogenesis in B16-F10 cells by inducing senescence. *Phytomedecine* **2018**, *46*, 1–10. [CrossRef]

30. Kubatka, P.; Kello, M.; Kajo, K.; Kruzliak, P.; Výbohová, D.; Mojžiš, J.; Adamkov, M.; Fialová, S.; Veizerová, L.; Zulli, A.; et al. Oregano demonstrates distinct tumor-suppressive effects in the breast carcinoma model. *Eur. J. Nutr.* **2017**, *56*, 1303–1316. [CrossRef]

31. Pezzani, R.; Vitalini, S.; Iriti, M. Bioactivities of *Origanum vulgare* L.: An update. *Phytochem. Rev.* **2017**, *16*, 1253–1268. [CrossRef]

32. Marrelli, M.; Statti, G.A.; Conforti, F. *Origanum* spp.: An update of their chemical and biological profiles. *Phytochem. Rev.* **2018**, *17*, 873. [CrossRef]

33. Kintzios, S.E. *In Handbook of Herbs and Spices*; Woodhead Publishing Ltd.: Abington Hall, Abington, MA, USA; Cambridge, UK, 2012; Volume 2, pp. 417–436.

34. Leyva-López, N.; Gutiérrez-Grijalva, E.P.; Vazquez-Olivo, G.; Heredia, J.B. Essential oils of oregano: Biological activity beyond their antimicrobial properties. *Molecules* **2017**, *22*, 989.

35. Oniga, I.; Pușcaș, C.; Silaghi-Dumitrescu, R.; Olah, N.K.; Sevastre, B.; Marica, R.; Marcus, I.; Sevastre-Berghian, A.C.; Benedec, D.; Pop, C.E.; et al. *Origanum vulgare* ssp. vulgare: Chemical composition and biological studies. *Molecules* **2018**, *23*, 2077. [CrossRef] [PubMed]

36. Impei, S.; Gismondi, A.; Canuti, L.; Canini, A. Metabolic and biological profile of autochthonous *Vitis vinifera* L. ecotypes. *Food Funct.* **2015**, *6*, 1526–1538. [CrossRef] [PubMed]

37. Chang, C.C.; Yang, M.H.; Wen, H.M.; Chern, J.C. Estimation of total flavonoid content in propolis by two complementary colorimetric methods. *J. Food Drug Anal.* **2002**, *10*, 3.

38. Fidler, I.J. Selection of successive tumor lines for metastasis. *Nat. New Biol.* **1973**, *242*, 148–149. [CrossRef]

39. Gismondi, A.; Lentini, A.; Tabolacci, C.; Provenzano, B.; Beninati, S. Transglutaminase-dependent antiproliferative and differentiative properties of nimesulide on B16-F10 mouse melanoma cells. *Amino Acids* **2010**, *38*, 257–262. [CrossRef]

40. Gismondi, A.; Canuti, L.; Impei, S.; Di Marco, G.; Kenzo, M.; Colizzi, V.; Canini, A. Antioxidant extracts of African medicinal plants induce cell cycle arrest and differentiation in B16F10 melanoma cells. *Int. J. Oncol.* **2013**, *43*, 956–964. [CrossRef]

41. Rossi, D.; Guerrini, A.; Maietti, S.; Bruni, R.; Paganetto, G.; Poli, F.; Scalvenzi, L.; Radice, M.; Saro, K.; Sacchetti, G. Chemical fingerprinting and bioactivity of Amazonian Ecuador *Croton lechleri* Müll. Arg. (Euphorbiaceae) stem bark essential oil: A new functional food ingredient? *Food Chem.* **2011**, *126*, 837–848. [CrossRef]

42. Rossi, D.; Guerrini, A.; Paganetto, G.; Bernacchia, G.; Conforti, F.; Statti, G.; Maietti, S.; Poppi, I.; Tacchini, M.; Sacchetti, G. *Croton lechleri* Müll. Arg. (Euphorbiaceae) stem bark essential oil as possible mutagen-protective food ingredient against heterocyclic amines from cooked food. *Food Chem.* **2013**, *139*, 439–447. [CrossRef] [PubMed]

43. Bradford, M.M. A rapid and sensitive method for the quantitation of microgram quantities of protein utilizing the principle of protein-dye binding. *Anal. Biochem.* **1976**, *72*, 248–254. [CrossRef]

44. Mansky, K.C.; Sulzbacher, S.; Purdom, G.; Nelsen, L.; Hume, D.A.; Rehli, M.; Ostrowski, M.C. The microphthalmia transcription factor and the related helix-loop-helix zipper factors TFE-3 and TFE-C collaborate to activate the tartrate-resistant acid phosphatase promoter. *J. Leukoc. Biol.* **2002**, *71*, 304–310.

45. Vetrini, F.; Auricchio, A.; Du, J.; Angeletti, B.; Fisher, D.E.; Ballabio, A.; Marigo, V. The microphthalmia transcription factor (Mitf) controls expression of the ocular albinism type 1 gene: Link between melanin synthesis and melanosome biogenesis. *Mol. Cell. Biol.* **2004**, *24*, 6550–6559. [CrossRef]

46. Zhang, Y.; Yuan, Y.; Liang, P.; Guo, X.; Ying, Y.; Shu, X.S.; Gao, M., Jr.; Cheng, Y. OSR1 is a novel epigenetic silenced tumor suppressor regulating invasion and proliferation in renal cell carcinoma. *Oncotarget* **2017**, *8*, 30008–30018. [CrossRef]

47. Lin, Y.T.; Lin, C.C.; Wang, H.C.; Hsu, Y.C. Induction of mitotic delay in pharyngeal and nasopharyngeal carcinoma cells using an aqueous extract of *Ajuga bracteosa*. *Int. J. Med. Sci.* **2017**, *14*, 462–469. [CrossRef] [PubMed]

48. Simard, M.J.; Chabot, B. Control of hnRNP A1 alternative splicing: An intron element represses use of the common 3′ splice site. *Mol. Cell. Biol.* **2000**, *20*, 7353–7362. [CrossRef] [PubMed]

49. Gismondi, A.; Canuti, L.; Grispo, M.; Canini, A. Biochemical composition and antioxidant properties of Lavandula angustifolia Miller essential oil are shielded by propolis against UV radiations. *Photochem. Photobiol.* **2014**, *90*, 702–708. [CrossRef] [PubMed]

50. Pendergrass, W.; Wolf, N.; Pool, M. Efficacy of MitoTracker GreenTM and CMXRosamine to measure changes in mitochondrial membrane potentials in living cells and tissues. *Cytom. A* **2004**, *61*, 162–169. [CrossRef]

51. Wani, M.C.; Taylor, H.L.; Wall, M.E.; Coggon, P.; Mcphail, A.T. Plant antitumor agents.VI. The isolation and structure of taxol, a novel antileukemic and antitumor agent from *Taxus brevifolia*. *J. Am. Chem. Soc.* **1971**, *93*, 2325–2327. [CrossRef] [PubMed]

52. Woods, C.M.; Zhu, J.; McQueney, P.A.; Bollag, D.; Lazarides, E. Taxol-induced mitotic block triggers rapid onset of a p53-independent apoptotic pathway. *Mol. Med.* **1995**, *5*, 506–526. [CrossRef]

53. Nurse, P. Universal control mechanism regulating onset of M-phase. *Nature* **1990**, *344*, 503–508. [CrossRef] [PubMed]

54. Coqueret, O. New roles for p21 and p27 cell-cycle inhibitors: A function for each cell compartment? *Trends Cell Biol.* **2003**, *13*, 65–70. [CrossRef]

55. Kastan, M.B.; Bartek, J. Cell-cycle checkpoints and cancer. *Nature* **2004**, *432*, 316–323. [CrossRef] [PubMed]

56. Massague, J. G1 cell-cycle control and cancer. *Nature* **2004**, *432*, 298–306. [CrossRef]

57. Tishler, R.B.; Lamppu, D.M.; Park, S.; Price, B.D. Microtubule-active drugs taxol, vinblastine, and nocodazole increase the levels of transcriptionally active p53. *Cancer Res.* **1995**, *55*, 6021–6025. [PubMed]

58. Maron, D.M.; Ames, B.N. Revised methods for the *Salmonella* mutagenicity test. *Mutat. Res.* **1983**, *113*, 173–215. [CrossRef]

59. Hartman, M.L.; Czyz, M. MITF in melanoma: Mechanisms behind its expression and activity. *Cell. Mol. Life Sci.* **2015**, *72*, 1249–1260. [CrossRef]

60. Slee, E.A.; Zhu, H.; Chow, S.C.; Macfarlane, M.; Nicholson, D.W.; Cohen, G.M. Benzyloxycarbonyl-Val-Ala-Asp (OMe) fluoromethylketone (Z-VAD.FMK) inhibits apoptosis by blocking the processing of CPP32. *Biochem. J.* **1996**, *315*, 21–24. [CrossRef] [PubMed]

61. Degterev, A.; Hitomi, J.; Germscheid, M.; Chen, I.L.; Korkina, O.; Teng, X.; Abbott, D.; Cuny, G.D.; Yuan, C.; Wagner, G.; et al. Identification of RIP1 kinase as a specific cellular target of necrostatins. *Nat. Chem. Biol.* **2008**, *4*, 313–321. [CrossRef]

62. Guo, C.; Sun, L.; Chen, X.; Zhang, D. Oxidative stress, mitochondrial damage and neurodegenerative diseases. *Neural Regen. Res.* **2013**, *8*, 2003–2014. [PubMed]

63. Carrì, M.T.; Valle, C.; Bozzo, F.; Cozzolino, M. Oxidative stress and mitochondrial damage: Importance in non-SOD1 ALS. *Front. Cell. Neurosci.* **2015**, *9*, 41. [CrossRef] [PubMed]

64. Whitaker-Menezes, D.; Martinez-Outschoorn, U.E.; Flomenberg, N.; Birbe, R.C.; Witkiewicz, A.K.; Howell, A.; Pavlides, S.; Tsirigos, A.; Ertel, A.; Pestell, R.G.; et al. Hyperactivation of oxidative mitochondrial metabolism in epithelial cancer cells in situ: Visualizing the therapeutic effects of metformin in tumor tissue. *Cell Cycle* **2011**, *10*, 4047–4064. [CrossRef] [PubMed]

65. Chikara, S.; Nagaprashantha, L.D.; Singhal, J.; Horne, D.; Awasthi, S.; Singhal, S.S. Oxidative stress and dietary phytochemicals: Role in cancer chemoprevention and treatment. *Cancer Lett.* **2018**, *413*, 122–134. [CrossRef] [PubMed]

66. Kuo, L.J.; Yang, L.X. Gamma-H2AX—A novel biomarker for DNA double-strand breaks. *In Vivo* **2008**, *22*, 305–309.

67. Schultz, L.B.; Chehab, N.H.; Malikzay, A.; Halazonetis, D. p53 binding protein 1 (53BP1) is an early participant in the cellular response to DNA double-strand breaks. *J. Cell Biol.* **2000**, *151*, 1381–1390. [CrossRef]

68. Said, A.M.; Fazal, F.; Rahman, A.; Hadi, S.M.; Parish, J.H. Activities of flavonoids for the cleavage of DNA in the presence of Cu(II): Correlation with the generation of active oxygen species. *Carcinogenesis* **1992**, *13*, 605–608.

69. The American Cancer Society. What are the Key Statistics about Melanoma Skin Cancer? 2017. Available online: http://www.cancer.org/cancer/skincancer-melanoma/detailedguide/melanoma-skin-cancer-key-statistics (accessed on 8 January 2020).

70. Narayanan, D.L.; Saladi, R.N.; Fox, J.L. Ultraviolet radiation and skin cancer. *Int. J. Dermatol.* **2010**, *49*, 978–986. [CrossRef]

71. Goldstein, A.M.; Chan, M.; Harland, M.; Hayward, N.K.; Demenais, F.; Bishop, D.T.; Azizi, E.; Bergman, W.; Bianchi-Scarra, G.; Bruno, W.; et al. Features associated with germline CDKN2A mutations: A GenoMEL study of melanoma-prone families from three continents. *J. Med. Genet.* **2007**, *44*, 99–106. [CrossRef]

72. Liu, Y.; Sheikh, M.S. Melanoma: Molecular pathogenesis and therapeutic management. *Mol. Cell. Pharmacol.* **2014**, *6*, 228.

73. Gismondi, A.; Di Marco, G.; Canuti, L.; Canini, A. Antiradical activity of phenolic metabolites extracted from grapes of white and red *Vitis vinifera* L. cultivars. *Vitis* **2017**, *56*, 19–26.

74. Jensen, J.D.; Wing, G.J.; Delvalle, R.P. Nutrition and melanoma prevention. *Clin. Dermatol.* **2010**, *28*, 644–649. [CrossRef] [PubMed]

75. Chinembiri, T.N.; du Plessis, L.H.; Gerber, M.; Hamman, J.H.; du Plessis, J. Review of natural compounds for potential skin cancer treatment. *Molecules* **2014**, *19*, 11679–11721. [CrossRef] [PubMed]

76. Tong, L.X.; Young, L.C. Nutrition: The future of melanoma prevention? *J. Am. Acad. Dermatol.* **2014**, *71*, 151–160. [CrossRef] [PubMed]

77. Działo, M.; Mierziak, J.; Korzun, U.; Preisner, M.; Szopa, J.; Kulma, A. The potential of plant phenolics in prevention and therapy of skin disorders. *Int. J. Mol. Sci.* **2016**, *17*, 160. [CrossRef] [PubMed]

78. Ettorre, A.; Frosali, S.; Andreassi, M.; Di Stefano, A. Lycopene phytocomplex, but not pure lycopene, is able to trigger apoptosis and improve the efficacy of photodynamic therapy in HL60 human leukemia cells. *Exp. Biol. Med.* **2010**, *235*, 1114–1125. [CrossRef] [PubMed]

79. Pietrocola, F.; Mariño, G.; Lissa, D.; Vacchelli, E.; Malik, S.A.; Niso-Santano, M.; Zamzami, N.; Galluzzi, L.; Maiuri, M.C.; Kroemer, G. Pro-autophagic polyphenols reduce the acetylation of cytoplasmic proteins. *Cell Cycle* **2012**, *11*, 3851–3860. [CrossRef] [PubMed]

80. Hasa, D.; Perissutti, B.; Dall'Acqua, S.; Chierotti, M.R.; Gobetto, R.; Grabnar, I.; Cepek, C.; Voinovich, D. Rationale of using *Vinca minor* Linne dry extract phytocomplex as a vincamine's oral bioavailability enhancer. *Eur. J. Pharm. Biopharm.* **2013**, *84*, 138–144. [CrossRef] [PubMed]

81. Kokkini, S.; Karousou, R.; Dardioti, A.; Krigas, N.; Lanaras, T. Autumn essential oils of greek oregano. *Phytochemistry* **1997**, *44*, 883–886. [CrossRef]

82. Russo, M.; Galletti, G.C.; Bocchini, P.; Carnacini, A. Essential oil chemical composition of wild populations of Italian oregano spice (*Origanum vulgare* ssp. hirtum (Link) Ietswaart): A preliminary evaluation of their use in chemotaxonomy by cluster analysis. 1. Inflorescences. *J. Agric. Food Chem.* **1998**, *46*, 3741–3746. [CrossRef]

83. De Martino, L.; De Feo, V.; Formisano, C.; Mignola, E.; Senatore, F. Chemical composition and antimicrobial activity of the essential oils from three chemotypes of *Origanum vulgare* L. ssp. hirtum (Link) Ietswaart growing wild in Campania (Southern Italy). *Molecules* **2009**, *14*, 2735–2746. [CrossRef] [PubMed]

84. Chun, S.S.; Vattem, D.A.; Lin, Y.T.; Shetty, K. Phenolic antioxidants from clonal oregano (*Origanum vulgare*) with antimicrobial activity against Helicobacter pylori. *Process Biochem.* **2005**, *40*, 809–816. [CrossRef]

85. Skendi, A.; Irakli, M.; Chatzopoulou, P. Analysis of phenolic compounds in Greek plants of Lamiaceae family by HPLC. *J. Appl. Res. Med. Aromat. Plants* **2017**, *6*, 62–69. [CrossRef]

86. Badisa, R.B.; Darling-Reed, S.F.; Joseph, P.; Cooperwood, J.S.; Latinwo, L.M.; Goodman, C.B. Selective cytotoxic activities of two novel synthetic drugs on human breast carcinoma MCF-7 cells. *Anticancer Res.* **2009**, *29*, 2993–2996. [PubMed]

87. EFSA Scientific Committee. Scientific Opinion on genotoxicity testing strategies applicable to food and feed safety assessment. *EFSA J.* **2011**, *9*, 2379.

88. Nurmi, A.; Mursu, J.; Nurmi, T.; Nyyssönen, K.; Alfthan, G.; Hiltunen, R.; Kaikkonen, J.; Salonen, J.T.; Voutilainen, S. Consumption of juice fortified with oregano extract markedly increases excretion of phenolic acids but lacks short- and long-term effects on lipid peroxidation in healthy nonsmoking men. *J. Agric. Food Chem.* **2006**, *54*, 5790–5796. [CrossRef] [PubMed]

89. Llana-Ruiz-Cabello, M.; Puerto, M.; Maisanaba, S.; Guzmán-Guillén, R.; Pichardo, S.; Cameán, A.M. Use of micronucleus and comet assay to evaluate evaluate the genotoxicity of oregano essential oil (*Origanum vulgare* L. *virens*) in rats orally exposed for 90 days. *J. Toxicol. Environ. Health A* **2018**, *81*, 525–533. [CrossRef]

90. Vaughn, C.J. Drugs and Lactation Database: LactMed. *J. Electron. Resour. Med. Libr.* **2012**, *9*, 272–277. [CrossRef]

91. EFSA Panel on Food Additives and Nutrient Sources added to Food (ANS). Scientific Opinion on the use of oregano and lemon balm extracts as a food additive on request of the European Commission. *EFSA J.* **2010**, *8*, 1514.

92. EFSA Scientific Committee. Technical report on the outcome of the consultation with Member States and EFSA on the basic substance application for *Origanum vulgare* L. essential oil for use in plant protection as fungicide, bactericide and insecticide. *EFSA Support. Publ.* **2016**, *13*, 1054.

93. Liang, C.H.; Chou, T.H.; Ding, H.Y. Inhibition of melanogensis by a novel origanoside from *Origanum vulgare*. *J. Dermatol. Sci.* **2010**, *57*, 170–177. [CrossRef] [PubMed]

94. Murphy, M.P. How mitochondria produce reactive oxygen species. *Biochem. J.* **2009**, *417*, 1–13. [CrossRef] [PubMed]

95. Arunasree, K.M. Anti-proliferative effects of carvacrol on a human metastatic breast cancer cell line, MDA-MB 231. *Phytomedicine* **2010**, *17*, 581–588. [CrossRef] [PubMed]

96. Khan, F.; Khan, I.; Farooqui, A.; Ansari, I. Carvacrol induces reactive oxygen species (ROS)-mediated apoptosis along with cell cycle arrest at G0/G1 in human prostate cancer cells. *Nutr. Cancer* **2017**, *69*, 1075–1087. [CrossRef] [PubMed]

97. Sharifi-Rad, M.; Varoni, E.M.; Iriti, M.; Martorell, M.; Setzer, W.N.; del Mar Contreras, M.; Salehi, B.; Soltani-Nejad, A.; Rajabi, S.; Tajbakhsh, M.; et al. Carvacrol and human health: A comprehensive review. *Phytother. Res.* **2018**, *32*, 1675–1687. [CrossRef] [PubMed]

98. Rodolfo, C.; Rocco, M.; Cattaneo, L.; Tartaglia, M.; Sassi, M.; Aducci, P.; Scaloni, A.; Camoni, L.; Marra, M. Ophiobolin a induces autophagy and activates the mitochondrial pathway of apoptosis in human melanoma cells. *PLoS ONE* **2016**, *11*, e0167672. [CrossRef]

99. Savini, I.; Arnone, R.; Catani, M.V.; Avigliano, L. *Origanum vulgare* induces apoptosis in human colon cancer caco2 cells. *Nutr. Cancer* **2009**, *61*, 381–389. [CrossRef]

100. Rubin, B.; Mansi, J.; Monticelli, H.; Bertazza, L.; Redaelli, M.; Sensi, F.; Zorzan, M.; Scaroni, C.; Mian, C.; Iacobone, M.; et al. Crude extract of *Origanum vulgare* L. induced cell death and suppressed MAPK and PI3/Akt signaling pathways in SW13 and H295R cell lines. *Nat. Prod. Res.* **2018**, *15*, 1646–1649. [CrossRef]

101. Christofferson, D.E.; Yuan, J. Necroptosis as an alternative form of programmed cell death. *Curr. Opin. Cell Biol.* **2010**, *22*, 263–268. [CrossRef]

102. Ashkenazi, A.; Salvesen, G. Regulated cell death: Signaling and mechanisms. *Annu. Rev. Cell Dev. Biol.* **2014**, *30*, 337–356. [CrossRef]

103. Sawai, H.; Domae, N. Discrimination between primary necrosis and apoptosis by necrostatin-1 in Annexin V-positive/propidium iodide-negative cells. *Biochem. Biophys. Res. Commun.* **2011**, *411*, 569–573. [CrossRef] [PubMed]

104. Al Dhaheri, Y.; Eid, A.; AbuQamar, S.; Attoub, S.; Khasawneh, M.; Aiche, G.; Hisaindee, S.; Iratni, R. Mitotic arrest and apoptosis in breast cancer cells induced by *Origanum majorana* extract: Upregulation of TNF-$\alpha$ and downregulation of survivin and mutant p53. *PLoS ONE* **2013**, *8*, e56649. [CrossRef] [PubMed]

105. Müller, M.; Mironov, S.L.; Ivannikov, M.V.; Schmidt, J.; Richter, D.W. Mitochondrial organization and motility probed by two-photon microscopy in cultured mouse brainstem neurons. *Exp. Cell Res.* **2005**, *303*, 114–127. [CrossRef]

106. Suen, D.F.; Norris, K.L.; Youle, R.J. Mitochondrial dynamics and apoptosis. *Genes Dev.* **2008**, *22*, 1577–1590. [CrossRef] [PubMed]

107. Gross, A.; McDonnell, J.M.; Korsmeyer, S.J. BCL-2 family members and the mitochondria in apoptosis. *Genes Dev.* **1999**, *13*, 1899–1911. [CrossRef] [PubMed]

108. Elmore, S. Apoptosis: A review of programmed cell death. *Toxicol. Pathol.* **2007**, *35*, 495–516. [CrossRef] [PubMed]

109. Lemasters, J.J. Selective Mitochondrial Autophagy, or Mitophagy, as a Targeted Defense against Oxidative Stress, Mitochondrial Dysfunction, and Aging. *Rejuvenation Res.* **2005**, *8*, 3–5. [CrossRef] [PubMed]

110. Lee, J.; Giordano, S.; Zhang, J. Autophagy, mitochondria and oxidative stress: Cross-talk and redox signalling. *Biochem. J.* **2012**, *441*, 523–540. [CrossRef]

111. Miyashita, T.; Reed, J.C. Tumor suppressor p53 is a direct transcriptional activator of the human bax gene. *Cell* **1995**, *80*, 293–299.

112. Hemann, M.T.; Lowe, S.W. The p53-Bcl-2 connection. *Cell Death Differ.* **2006**, *13*, 1256–1259. [CrossRef]

113. Bourgarel-Rey, V.; Savry, A.; Hua, G.; Carré, M.; Bressin, C.; Chacon, C.; Imbert, J.; Braguer, D.; Barra, Y. Transcriptional down-regulation of Bcl-2 by vinorelbine: Identification of a novel binding site of p53 on Bcl-2 promoter. *Biochem. Pharmacol.* **2009**, *78*, 1148–1156. [CrossRef] [PubMed]

114. Shiloh, Y.; Ziv, Y. The ATM protein kinase: Regulating the cellular response to genotoxic stress, and more. *Nat. Rev. Mol. Cell Biol.* **2013**, *14*, 197–210. [CrossRef] [PubMed]

115. Williams, A.B.; Schumacher, B. p53 in the DNA-damage-repair process. *Cold Spring Harb. Perspect. Med.* **2016**, *6*, a026070. [CrossRef]

116. Kagawa, T.F.; Geierstanger, B.H.; Wang, A.H.J.; Ho, P.S. Covalent modification of guanine bases in double-stranded DNA. The 1.2-A Z-DNA structure of d(CGCGCG) in the presence of $CuCl_2$. *J. Biol. Chem.* **1991**, *266*, 20175–20184. [PubMed]

117. Ullah, M.F.; Ahmad, A.; Khan, H.Y.; Zubair, H.; Sarkar, F.H.; Hadi, S.M. The prooxidant action of dietary antioxidants leading to cellular DNA breakage and anticancer effects: Implications for chemotherapeutic action against cancer. *Cell Biochem. Biophys.* **2013**, *67*, 431–438. [CrossRef] [PubMed]

118. Narciso, L.; Fortini, P.; Pajalunga, D.; Franchitto, A.; Liu, P.; Degan, P.; Frechet, M.; Demple, B.; Crescenzi, M.; Dogliotti, E. Terminally differentiated muscle cells are defective in base excision DNA repair and hypersensitive to oxygen injury. *Proc. Natl. Acad. Sci. USA* **2007**, *104*, 17010–17015. [CrossRef] [PubMed]

119. Kinner, A.; Wu, W.; Staudt, C.; Iliakis, G. Gamma-H2AX in recognition and signaling of DNA double-strand breaks in the context of chromatin. *Nucleic Acids Res.* **2008**, *36*, 5678–5694. [CrossRef] [PubMed]

120. Cleaver, J.E.; Feeney, L.; Revet, I. Phosphorylated H2Ax is not an unambiguous marker for DNA double strand breaks. *Cell Cycle* **2011**, *10*, 3223–3224. [CrossRef] [PubMed]

121. Fortini, P.; Ferretti, C.; Pascucci, B.; Narciso, L.; Pajalunga, D.; Puggioni, E.M.R.; Castino, R.; Isidoro, C.; Crescenzi, M.; Dogliotti, E. DNA damage response by single-strand breaks in terminally differentiated muscle cells and the control of muscle integrity. *Cell Death Differ.* **2012**, *19*, 1741–1749. [CrossRef] [PubMed]

122. Ward, I.M.; Minn, K.; van Deursen, J.; Chen, J. p53 binding protein 53BP1 is required for DNA damage responses and tumor suppression in mice. *Mol. Cell. Biol.* **2003**, *23*, 2556–2563. [CrossRef]

123. Dearfield, K.L.; Cimino, M.C.; Mccarroll, N.E.; Mauer, I.; Valcovic, L.R. Genotoxicity risk assessment: A proposed classification strategy. *Mutat. Res.* **2002**, *521*, 121–135. [CrossRef]

124. Laughton, M.J.; Halliwell, B.; Evans, P.J.; Robin, J.; Hoult, S. Antioxidant and pro-oxidant actions of the plant phenolics quercetin, gossypol and myricetin. Effects on lipid peroxidation, hydroxyl radical generation and bleomycin-dependent damage to DNA. *Biochem. Pharmacol.* **1989**, *38*, 2859–2865. [CrossRef]

125. Hadi, S.M.; Bhat, S.H.; Azmi, A.S.; Hanif, S.; Shamim, U.; Ullah, M.F. Oxidative breakage of cellular DNA by plant polyphenols: A putative mechanism for anticancer properties. *Semin. Cancer Biol.* **2007**, *17*, 370–376. [CrossRef] [PubMed]

126. Tsai, Y.C.; Wang, Y.H.; Liou, C.C.; Lin, Y.C.; Huang, H.; Liu, Y.C. Induction of oxidative DNA damage by flavonoids of propolis: Its mechanism and implication about antioxidant capacity. *Chem. Res. Toxicol.* **2012**, *25*, 191–196. [CrossRef] [PubMed]

127. Ebadi, M.; Swanson, S. The status of zinc, copper, and metallothionein in cancer patients. *Prog. Clin. Biol. Res.* **1988**, *259*, 161–175. [PubMed]

128. Yoshida, D.; Ikeda, Y.; Nakazawa, S. Quantitative analysis of copper, zinc and copper/zinc ratio in selected human brain tumors. *J. Neurooncol.* **1993**, *16*, 109–115. [CrossRef] [PubMed]

129. Gupte, A.; Mumper, R.J. Elevated copper and oxidative stress in cancer cells as a target for cancer treatment. *Cancer Treat. Rev.* **2009**, *35*, 32–46. [CrossRef] [PubMed]

130. Khan, H.Y.; Zubair, H.; Ullah, M.F.; Ahmad, A.; Hadi, S.M. A prooxidant mechanism for the anticancer and chemopreventive properties of plant polyphenols. *Curr. Drug Targets* **2012**, *13*, 1738–1749. [CrossRef] [PubMed]

131. Denoyer, D.; Masaldan, S.; La Fontaine, S.; Cater, M.A. Targeting copper in cancer therapy: "Copper That Cancer". *Metallomics* **2015**, *7*, 1459–1476. [CrossRef] [PubMed]

132. Garber, K. Cancer's copper connections. *Science* **2015**, *349*, 129. [CrossRef]

133. Sergediene, E.; Jönsson, K.; Szymusiak, H.; Tyrakowska, B.; Rietjens, I.M.C.M.; Čenas, N. Prooxidant toxicity of polyphenolic antioxidants to HL-60 cells: Description of quantitative structure-activity relationships. *FEBS Lett.* **1999**, *462*, 392–396. [CrossRef]

134. Chen, Q.; Espey, M.G.; Krishna, M.C.; Mitchell, J.B.; Corpe, C.P.; Buettner, G.R.; Shacter, E.; Levine, M. Pharmacologic ascorbic acid concentrations selectively kill cancer cells: Action as a pro-drug to deliver hydrogen peroxide to tissues. *Proc. Natl. Acad. Sci. USA* **2005**, *102*, 13604–13609. [CrossRef] [PubMed]

135. Taniguchi, H.; Yoshida, T.; Horinaka, M.; Yasuda, T.; Goda, A.E.; Konishi, M.; Wakada, M.; Kataoka, K.; Yoshikawa, T.; Sakai, T. Baicalein overcomes tumor necrosis factor-related apoptosis-inducing ligand resistance via two different cell-specific pathways in cancer cells but not in normal cells. *Cancer Res.* **2008**, *68*, 8918–8927. [CrossRef] [PubMed]

136. Wei, L.; Lu, N.; Dai, Q.; Rong, J.; Chen, Y.; Li, Z.; You, Q.; Guo, Q. Different apoptotic effects of wogonin via induction of $H_2O_2$ generation and $Ca^{2+}$ overload in malignant hepatoma and normal hepatic cells. *J. Cell. Biochem.* **2010**, *111*, 1629–1641. [CrossRef] [PubMed]

137. Wilken, R.; Veena, M.S.; Wang, M.B.; Srivatsan, E.S. Curcumin: A review of anti-cancer properties and therapeutic activity in head and neck squamous cell carcinoma. *Mol. Cancer* **2011**, *10*, 12. [CrossRef] [PubMed]

138. Calderón-Montaño, J.M.; Martínez-Sánchez, S.M.; Burgos-Morón, E.; Guillén-Mancina, E.; Jiménez-Alonso, J.J.; García, F.; Aparicio, A.; López-Lázaro, M. Screening for selective anticancer activity of plants from Grazalema Natural Park, Spain. *Nat. Prod. Res.* **2018**, *33*, 3454–3458. [CrossRef] [PubMed]

139. Cheimonidi, C.; Samara, P.; Polychronopoulos, P.; Tsakiri, E.N.; Nikou, T.; Myrianthopoulos, V.; Sakellaropoulos, T.; Zoumpourlis, V.; Mikros, E.; Papassideri, I.; et al. Selective cytotoxicity of the herbal substance acteoside against tumor cells and its mechanistic insights. *Redox. Biol.* **2018**, *16*, 169–178. [CrossRef]

**Publisher's Note:** MDPI stays neutral with regard to jurisdictional claims in published maps and institutional affiliations.

Communication

# Antioxidant Capacity of Tempura Deep-Fried Products Prepared Using Barley, Buckwheat, and Job's Tears Flours

Asuka Taniguchi [1], Nami Kyogoku [1,2], Hiroko Kimura [3], Tsubasa Kondo [3], Keiko Nagao [1] and Rie Kobayashi [1,3,*]

[1] Graduate School of Humanities and Life Sciences, Tokyo Kasei University, Tokyo 1738602, Japan; g190301@tokyo-kasei.ac.jp (A.T.); kyogoku@kanazawa-gu.ac.jp (N.K.); nagao@tokyo-kasei.ac.jp (K.N.)
[2] Department of Food and Nutrition, Kanazawa Gakuin College, Ishikawa 9201392, Japan
[3] Faculty of Home Economics, Tokyo Kasei University, Tokyo 1738602, Japan; 716hiroko.w@gmail.com (H.K.); kondo-t@tokyo-kasei.ac.jp (T.K.)
* Correspondence: kobayashi-r@tokyo-kasei.ac.jp; Tel.: +81-3-3961-7248

Received: 7 August 2020; Accepted: 2 September 2020; Published: 7 September 2020

**Abstract:** Tempura is a dish of battered and deep-fried foods, and wheat flour is typically used; however, barley, buckwheat, and Job's tears have an antioxidant capacity. This study investigated whether replacing wheat flour with flours from these three crops in tempura affects the antioxidant capacity and deterioration of frying oil. Radical scavenging activity and polyphenol content of tempura were measured by chemiluminescence-based assay and the Folin–Denis method, respectively. The peroxide value, $p$-anisidin value, acid value, and polar compound of the oil used in frying were measured as indexes of oil deterioration post-frying due to oxidation. Although the frying oil of barley showed higher $p$-anisidin value than that of wheat, the oil samples' deterioration level measured in this study was low. The antioxidant capacity and polyphenol content in the three flours samples were higher than those in wheat sample, with buckwheat producing the greatest values, followed by Job's tears, and then barley. Thus, deep-fried products prepared using the three flours demonstrated superior antioxidant capacity owing to the abundance of antioxidant components. Therefore, tempura can be enjoyed in a healthier manner by using batter prepared using those flours, and substituting wheat flour with the three flours can increase the antioxidant capacity of deep-fried products.

**Keywords:** tempura; deep-fried product; barley; buckwheat; Job's tears; antioxidant capacity; oil deterioration; polyphenol

## 1. Introduction

Tempura is a traditional Japanese dish that has gained popularity all over the world. It is prepared from a batter of water and flour; sea food or vegetable is lightly dipped in the batter and then instantly fried [1]. In general, wheat flour is used to prepare tempura; however, the structural characteristics of gluten present in wheat flour disturb the process of cooking high-quality tempura. The quality improvement of tempura has been studied: Nakamura et al. [2] used rice flour in their experiment; Matsunaga et al. [1] added different types of starch from various plants to wheat proteins, and Carvalho et al. [3] replaced water during batter preparation with ethanol and a $CO_2$ injection. Each of the methods described above effectively improved the quality of tempura. Another study by Martínez-Pineda et al. [4] prepared tempura using batter with malt dextrin, ethanol, baking powder, and corn flour. They concluded that adapting batter ingredients according to the purpose was an effective strategy to improve the quality of deep-fried products. Despite these studies conducted on tempura quality, no research has examined the use of barley, buckwheat, and Job's tears that do not construct gluten structure in preparing tempura.

In addition, these flours have several health benefits [5–7]. Regular intake of these grains reduces the risk of several chronic diseases associated with oxidative stress. Therefore, the biofunctionality of these flours is an attractive option for people who are looking for healthier options. Studies have demonstrated that these grains have high antioxidant capacities [8,9]. However, few studies have focused on the antioxidant capacity of these flours in deep-fried products, especially tempura.

We hypothesized that these flours of barley, buckwheat, and Job's tears not only provide health benefits but also make cooking tempura easier and improve its texture. We hope that research on tempura prepared using these flours will help establish their wider use for these flours. The quality and palatability of tempura prepared using those flours have already been evaluated in our previous study [10]. Hence, this study examined the antioxidant capacity of tempura prepared using barley, buckwheat, and Job's tears flours.

In the human body, the antioxidant activity of foods sequesters reactive oxygen is generated by electron reduction during metabolism [11]. Usually, reactive oxygen acts on immune functions [12], but when these oxygen groups are over produced, they cause lifestyle-related disease. Although the human body has antioxidant enzymes to extinguish reactive oxygen [12], they are not enough to inhibit excess active oxygen. The peroxyl radical, which is secondarily generated from the oxidation of cell lipids by active oxygen generated with energy production, is a particularly toxic reactive oxygen species, which promotes serial oxidation and spreads cell disorder. To erase peroxyl radicals, which cannot be absorbed or sequestered by other means in the human body, antioxidant activity in foods is essential. Antioxidant components derived from foods for peroxyl radicals are expected to reduce the risks of lifestyle-related diseases.

In addition, it is necessary to manage the quality of oil as tempura is a dish in which the ingredients are deep-fried in oil. In Japan, there are standards [13] for foods that use a lot of oil during cooking to manage oil quality. One of the rules has to do with peroxide value (PV) and acid value (AV). These are indexes of fat deterioration in foods. Oil deterioration causes a change in smell, taste, and color and decreases nutritive value. Deteriorated oil gives detrimental effects, such as nausea, abdominal pain, and lassitude to the human body. Thus, deterioration of oil is an important index and is related to health function. Since the deterioration mechanism of heated oil is complicated and its evaluation is necessary using multiple indicators, *p*-anisidin value (AnV) and polar compound (PC) were also examined.

Therefore, the present study investigated whether the use of three flours of barley, buckwheat, and Job's tears in preparing tempura could be useful for increased antioxidant capacity as a new added value. In this analysis, we evaluated the antioxidant capacity of their tempura based on peroxyl radical scavenging activity, polyphenol content, and oil oxidation.

## 2. Materials and Methods

### 2.1. Ingredients

Barley flour (Fiber snow; six-row barley, JA Komatsushi), buckwheat flour (common buckwheat, Nikkoku Seifun Corp., Nagano, Japan), and Job's tears flour (Akishizuku, Takada Hiryo–ten Co., Tochigi, Japan) were used. In addition to these flours, weak wheat flour (Hoshihime, protein content 8.8 g, Nikkoku Seifun Corp., Nagano, Japan) was used to compare with the three flours.

### 2.2. Preparation of Tempura Samples

The water content ratio for tempura is usually 160% of the weight of the wheat flour. Water ratios for the barley, buckwheat, and Job's tears flours were adjusted until the viscosity based on the steady flow measured using a rheometer (HAAKE MARS II, Thermo Fisher Scientific, Inc., Waltham, MA, USA) of the resulting batters was consistent with that of the wheat-based batter [14]. These ratios were determined to be as follows: barley, 260%; buckwheat, 190%; Job's tears, 190%. After adding water to each flour, they were stirred 60 times at a speed of once every second. The buckwheat and Job's tears batter were left to rest for 30 min, and then they were stirred again 10 times. The four batters were dispensed into silicon cups ($\varphi$30 mm, height 10 mm, Daiso Industries Co., Ltd., Hiroshima, Japan) by 2.0 mL each and fried in canola oil (stored unopened in a dark and cool place, The Nisshin OilliO Group, Ltd., Tokyo, Japan) for 140 s at a temperature of 180 °C–190 °C using a fryer (CDF-100, Cuisinart Compact Deep Fryer, Cuisinart, Stamford, CT, USA). After cooling for 1 min, they were preserved in a freezer (MDF-300E1, deep freezer, Fukushima Industries Corp., Osaka, Japan) set at −80 °C for at least 24 h. In the following measurements, six tempura samples were milled and analyzed in each test.

## 2.3. Preparation of Oil after Frying Samples

After frying, the oil was used for PV, AV, AnV and PC analysis. Frying oil samples were heated for 3 h at 180 °C ± 5 °C. During this time, 48 samples of tempura were fried and allowed to radiate heat naturally after heating (Figure 1).

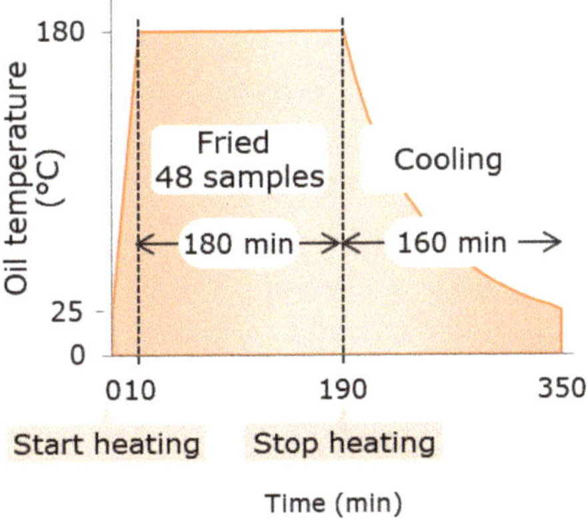


**Figure 1.** Tempura frying process used in this study.

## 2.4. Chemiluminescence Method

After preliminary freezing, the samples were freeze-dried (DC800, Freeze Dryer, Yamato Scientific Co., Ltd., Tokyo, Japan) and then crushed using a mill (IFM-800, Iwatani Corp., Osaka, Japan). Crushed samples were used for analysis of chemiluminescence and measurement of peroxyl radical scavenging activity, as described previously [15]. The results were measured using a photon counter (Lumitester C-100, Kikkoman Co., Ltd., Chiba, Japan) and shown as $IC_{50}$ values. $IC_{50}$ was defined as the sample density that was half the light emission value of the phosphate buffer solution without an antioxidant component. This was calculated as previously reported [15], and the results were expressed in Trolox equivalents.

## 2.5. Analysis of Oil Oxidation

Frying oils include various components, such as peroxides, which are produced by primary oxidation; carbonyl compounds, which are secondary products of this oxidation; free fatty acids, which are produced by hydrolysis, and polymerized glycerols. In the present study, PV, AV, AnV, and PC were measured to evaluate the complicated deterioration of frying oils. The AV, AnV, and PC were measured according to the standard methods of the Japan Oil Chemists' Society [16].

PV was measured based on previous standards [13]. Sample oil (5 g) was mixed with 10 mL of chloroform (FUJIFILM Wako Pure Chemical Corp., Tokyo, Japan) and 15 mL of glacial acetic acid. Potassium iodide saturated solution (1 mL; FUJIFILM Wako Pure Chemical Corp.) was added, which was prepared with 70 g of potassium iodide dissolved (FUJIFILM Wako Pure Chemical Corp.) in 50 mL of boiled deionized water and shaken intensely for 1 min. They were left for 10 min in the dark, and after the addition of 75 mL of deionized water, the samples were again intensely shaken.

Subsequently, they underwent titrimetric analysis with 0.01 M sodium thiosulfate (FUJIFILM Wako Pure Chemical Corp.), which is a standard solution, and 1% starch solution, which is an indicator that 1 g of soluble starch (FUJIFILM Wako Pure Chemical Corp.) was boiled and dissolved with 100 mL of deionized water. The 0.01 M sodium thiosulfate was diluted from 0.1 M sodium thiosulfate that had been previously quantified. This titration comprised the blank containing no oil. The formula to calculate PV was as follows:

$$PV = 0.01 \times (A - B) \times F \times 1000/S \tag{1}$$

where A is the titration amount of 0.01 M sodium thiosulfate in the main examination (mL), B is the titration amount of 0.01 M sodium thiosulfate in the blank examination (mL), F is the titer of 0.1 M sodium thiosulfate, and S is the weight of the oil samples (g).

*2.6. Polyphenol Contents*

Total polyphenol content in each flour type and tempura sample were measured by the Folin–Denis method [17]. Each flour type or sample (1.0 g) was mixed with 80% ethanol/water solution and heated to 80 °C while stirring with a magnetic stirrer. After 30 min, they were centrifuged at 3000 $\times g$ for 10 min and then treated as extracted samples. A 1.0 mL aliquot from each extracted sample was mixed with 1.0 mL of Folin & Ciocalteu's Phenol reagent (MP Biomedicals, Inc., Santa Ana, USA) and left for 3 min. Sodium bicarbonate solution (1.0 mL, 10% w/w) was added, stirred, and reacted for 60 min at a room temperature. The absorbance at 700 nm was measured using a spectrophotometer (UH5300, HITACHI Ltd., Tokyo, Japan). Results were expressed as mg of catechin per 100 g of sample.

*2.7. Statistical Analysis*

Means ± standard deviation was calculated and analyzed for significant differences by Tukey's test of multiple comparisons after performing analysis of variance using R software (3.3.4). For samples, values of $p < 0.05$ were considered significant.

## 3. Results

*3.1. Antioxidant Capacity of Barley, Buckwheat, and Job's Tears Flours and Their Tempura*

Peroxyl radical scavenging activity was measured and expressed in Trolox equivalents. In this unit, a higher value indicates a higher antioxidant capacity (Figure 2). The antioxidant values of the flours of barley, buckwheat, and Job's tears were significantly higher than those of wheat flour. When comparing the flour and the fried batter in barley, buckwheat, and Job's tears, the antioxidant values of fried batter were significantly lower than that of the flour. However, the antioxidant values of buckwheat and the Job's tears samples were higher than that in the wheat sample also in fried batter. Although the average value of the radical scavenging activity was higher in the barley sample than in the wheat sample, there was no significant difference between the barley and wheat samples.

*3.2. Oxidation of the Oil Used in Frying*

In this study, PV, AV, AnV, and PC were measured to evaluate the oxidation of the oil used for frying.

AV is the concentration free fatty acids in the oil due to the hydrolysis of oil [18], and PC measures oxidization, dimerization, and polymerization of triacylglycerols and diacylglycerols and free fatty acids [19]. AV was 0.1 for all sample oils, while PC was as follows: wheat, 6.0% ± 0.24%; barley, 5.7% ± 0.42%; buckwheat, 5.4% ± 0.12%; Job's tears, 5.6% ± 0.26%. These indexes of frying oils were at the same level, and there was no significant difference.

PV and AnV were compared, and the results are showed in Table 1. PV is an index of peroxides; however, since peroxides are thermally unstable and decompose at high temperatures [20], AnV that evaluates its decomposition products, aldehydes, and ketones was examined. The oil samples used for frying barley, buckwheat, and Job's tears flours had a lower PV than that used for wheat flour. Although the oil used for buckwheat and Job's tears samples did not show significant difference compared with the oil used for wheat and barley sample, AnV of the oil used for barley sample was higher than that used for wheat sample.

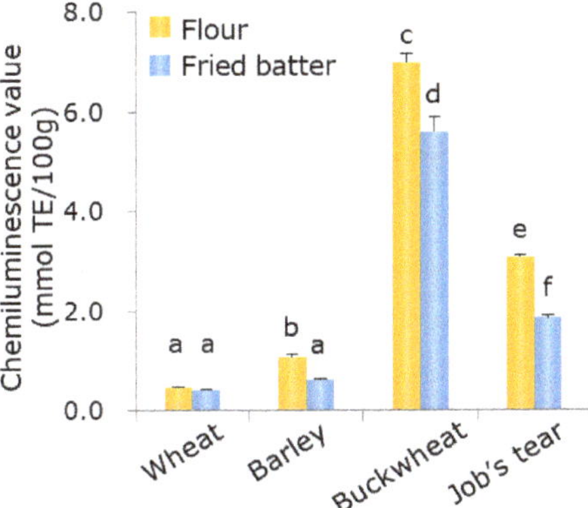


**Figure 2.** Peroxyl radical scavenging capacity in the four flours examined in this study. The $IC_{50}$ value was defined as the sample density that was half the value of the phosphate buffer solution wherein light emissions did not include an antioxidant component. Results are expressed as Trolox equivalents and measured using a chemiluminescence-based assay. Each value represents the mean ± SD ($n = 3$). Different letters a-f for flours and fried batter indicate a significant difference ($p < 0.05$, Tukey test).

**Table 1.** Peroxide value (PV) and *p*-anisidin value (AnV) of oil after frying in the four flours examined in this study.

|  | PV | AnV |
|---|---|---|
| Wheat | 4.82 ± 1.66 [a] | 39.3 ± 1.98 [a] |
| Barley | 2.61 ± 1.22 [b] | 53.2 ± 6.58 [b] |
| Buckwheat | 3.02 ± 1.26 [b] | 43.8 ± 1.30 [a,b] |
| Job's Tears | 3.15 ± 1.71 [b] | 45.6 ± 3.00 [a,b] |

The index of oil deterioration of the oil used to fry each of the 48 samples was measured to evaluate primary oxidation. Each value is expressed as the mean ± SD ($n = 3$). Different letters a-b indicate a significant difference ($p < 0.05$, Tukey test).

## 3.3. Polyphenol Content

As with antioxidant capacity, barley, buckwheat, and Job's tears flours displayed higher polyphenol content than wheat flour (Figure 3). Although the polyphenol content was significantly decreased by frying, tempura samples prepared using buckwheat and Job's tears flours maintained high levels of polyphenol. The polyphenol content of barley sample was between that of wheat and Job's tears samples, with no significant difference. The polyphenol content in tempura was highest for buckwheat, followed by Job's tears, barley, and finally wheat. This was the same order as that of antioxidant capacity.

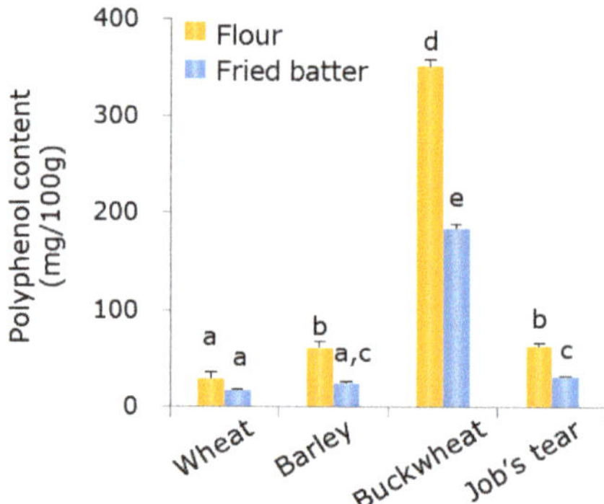


**Figure 3.** Polyphenol content in each of the flours examined in this study, measured using the Folin–Denis method. Each value represents the mean ± SD ($n$ = 3). Different letters a-e for flours and fried batter indicate a significant difference ($p$ < 0.05, Tukey test).

## 4. Discussion

In this study, we investigated whether substituting wheat flour with flours of barley, buckwheat, and Job's tears in deep-fried batter enhances tempura's antioxidant capacity and suppresses deterioration of the frying oil. In addition, we also examined the contribution of polyphenol present in those grains to these effects.

It was confirmed that the peroxyl radical scavenging activity in barley, buckwheat, and Job's tears flours was significantly higher than that of wheat flour. Among the three flours, buckwheat and Job's tears samples had high peroxyl radical scavenging activity, even after deep-frying. In particular, the buckwheat sample had the highest antioxidant value among the four flours examined, comparable with that of ginger [21], which is known to possess antioxidant activity [22]. Although other studies have reported that the use of these flours in baked products increases the antioxidant capacity [23,24], this study found that similar effects could be achieved with deep-fried products using these flours.

In the analysis of the deterioration process of frying oil, the PV and AnV of each sample showed significant differences, and there was a moderate negative correlation between the parameters at $r$ = 0.614 (Pearson product-moment correlation coefficient, $p$ < 0.05). A high AnV signifies that the decomposition of primary oxidation products is progressing; i.e., the oil in which the samples of barley flour, buckwheat flour, and Job's tears flour were fried tended to proceed to secondary oxidation compared with the wheat sample. However, AV and PC were low in all sample oils, indicating that the oil deterioration did not progress. The AV was extremely low, and limited free fatty acid was produced during frying. In Europe, the PC is used to regulate the oil used for frying [25]. According to the regulations, frying oil with 25%–27% PC should be changed, and the PC of the frying oil samples in this study were 5.6%–6.0% lower than the specified range in the regulation. We suggest that the oil samples used in this study began to deteriorate; in some of the oil samples, we observed foaming, which indicates deterioration. Nevertheless, as mentioned above, not all the oil samples deteriorated in the frying process.

Deteriorated oil causes oxidative rancidity and taste deterioration, and it contains components that are harmful to the body. These components, which are generated by the deterioration of the oil, are absorbed in deep-fried products with the frying oil and reduce the quality of the oil used for frying as well as the deep-fried products. The toxicity of the deteriorated oil has been studied, and it has been suggested that the cause of toxic effects in heating oil is due to polymerization [26]. In this study, the deterioration level of the oil, including the polymer, was evaluated using PC. However, the PC of the three samples was not higher than that of the wheat flour. It is presumed that the risk of health damage did not increase through the use of these three flours. However, the deterioration of oil in this study was extremely low, and it was not at a level to judge the harmful effects on the human body.

Although there was no evident difference in the deterioration between the frying oil samples in the results of this study, it may be necessary to increase the number of frying samples and the heating time to evaluate the deterioration level of each oil sample over time.

Barley, buckwheat, and Job's tears contain antioxidant components [27–29]. Polyphenol was one of the components that was measured in this study. The flours of these samples, especially buckwheat, had much higher polyphenol content than the wheat flour and the amounts thereof ranked in the same order as those for the radical scavenging assay. In addition, the correlation between the radical scavenging activity and polyphenol content was $r_s = 0.953$ (Spearman's rank correlation coefficient, $p < 0.001$), suggesting that polyphenol in those flours is an active antioxidant that contributes to the high antioxidant capacity of tempura, which is fried at a high temperature.

Therefore, the fried samples using the three flours, particularly the buckwheat and Job's tears flour samples, showed high peroxyl radical scavenging capacity as well as polyphenol content, and the use of barley, buckwheat, and Job's tears flours for cooking may result in tempura containing a higher antioxidant capacity than that prepared using wheat flour.

People on dietary restrictions, such as caloric control, tend to avoid tempura because it is a deep-fried product. We previously examined the quality and palatability of tempura using flours of barley, buckwheat, and Job's tears and found that the odor indices of those samples decreased compared with that of raw batter of the three flours and that oil absorption was approximately the same as or lower than that of the wheat sample. In the preference sensory evaluations, those samples received good evaluation, except for the grayish color of the buckwheat sample (Figure 4) [10]. Although it has already been evaluated that devising a suitable preparation method for tempura is effective in improving its texture and suppressing oil absorption [1–4], our results suggest that the use of barley, buckwheat, and Job's tears is a useful substitute for wheat flour when cooking tempura because it maintains its favorable characteristics, such as crispy texture, fragrant smell, and reduction in oil absorption. Moreover, the present study demonstrated that the use of those flours increases the antioxidant capacity of tempura as a new added value. Therefore, the use of barley, buckwheat, and Job's tears flours is expected to allow more people to enjoy tempura in a healthier manner. The finding that deep-fried products can acquire an antioxidant capacity by substituting wheat flour with barley, buckwheat, and Job's tears flours is a significant benefit that could be applicable to other deep-fried foods as well.

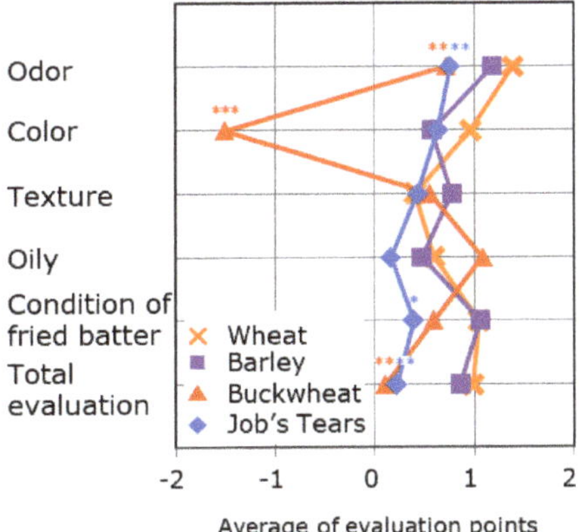



**Figure 4.** Preference sensory evaluation. Each value is expressed as the mean for each criterion. *** $p < 0.001$, ** $p < 0.01$, * $p < 0.05$ compared with the wheat sample (Dunnett test), $n = 49$. Panel: female university students and teachers aged 18–26 years. Evaluation method: 5-point scoring method of −2 (least preferred) to 2 (preferred). Evaluation sample: sweet potato tempura using each flour examined in this study in the batter. These data were previously reported in [10].

**Author Contributions:** Conceptualization, A.T. and R.K.; investigation, A.T., N.K., H.K., and T.K.; writing—original draft preparation, A.T.; writing—review and editing, A.T., K.N., and R.K.; project administration, R.K.; funding acquisition, R.K. All authors have read and agreed to the published version of the manuscript.

**Funding:** This research was funded by Research Institute of Domestic Science of Tokyo Kasei University and THE TOYO SUISAN FOUNDATION.

**Acknowledgments:** The authors would like to thank Nikkoku Seifun Corp. and JA Komatsu–shi for supplying sample flours used in the study.

**Conflicts of Interest:** The authors declare no conflict of interest.

## References

1. Matsunaga, K.; Kawasaki, S.; Takeda, Y. Influence of physicochemical properties of starch on crispness of Tempura fried batter. *Cereal Chem.* **2003**, *80*, 339–345. [CrossRef]

2. Nakamura, S.; Ohtsubo, K. Influence of physicochemical properties of rice flour on oil uptake of tempura frying batter. *Biosci. Biotechnol. Biochem.* **2010**, *74*, 2484–2489. [CrossRef] [PubMed]

3. Carvalho, M.J.; Ruiz-Carrascal, J. Improving crunchiness and crispness of fried squid rings through innovative tempura coatings: Addition of alcohol and $CO_2$ incubation. *J. Food Sci. Technol.* **2018**, *55*, 2068–2078. [CrossRef]

4. Martínez-Pineda, M.; Yagüe-Ruiz, C.; Vercet, A. How batter formulation can modify fried tempura-battered zucchini chemical and sensory characteristics? *Foods* **2020**, *9*, 626. [CrossRef] [PubMed]

5. Shimizu, C.; Kihara, M.; Aoe, S.; Araki, S.; Ito, K.; Hayashi, K.; Watari, J.; Sakata, Y.; Ikegami, S. Effect of high β–glucan barley on serum cholesterol concentrations and visceral fat area in Japanese men—A randomized, double–blinded, placebo–controlled trial. *Plant Foods Hum. Nutr.* **2008**, *63*, 21–25. [CrossRef]

6. Giménez–Bastida, A.; Zieliński, H. Buckwheat as a functional food and its effects on health. *J. Agric. Food Chem.* **2015**, *63*, 7896–7913. [CrossRef]

7. Ukita, T.; Tanimura, A. Studies on the anti–tumor component in the seeds of Coix Lachryma–Jobi L. VAR. Ma–yuen (ROMAN.) STAPF. I.: Isolation and anti–tumor activity of coixenolide. *Chem. Pharm. Bull.* **1961**, *9*, 43–46. [CrossRef]

8. Morishita, T.; Yamaguchi, H.; Degi, K. The contribution of polyphenols to antioxidative activity in common buckwheat and tartary buckwheat grain. *Plant Prod. Sci.* **2007**, *10*, 99–104. [CrossRef]

9. Adom, K.K.; Liu, R.H. Antioxidant activity of grains. *J. Agric. Food Chem.* **2002**, *50*, 6182–6187. [CrossRef]

10. Taniguchi, A.; Maruyama, R.; Kyogoku, N.; Nagao, K.; Kobayashi, R. Evaluation of suitability of millet flour in tempura batter based on quality and palatability. *J. Cook. Sci. Jpn.* **2019**, *52*, 376–385. (In Japanese)

11. Yoshikawa, T.; Naito, Y. What is oxidative stress? *Jpn. Med. Assoc. J.* **2002**, *45*, 271–276.

12. Kalyanaraman, B. Teaching the basics of redox biology to medical and graduate students: Oxidants, antioxidants and disease mechanisms. *Redox Biol.* **2013**, *1*, 244–257. [CrossRef] [PubMed]

13. Ministry of Health and Welfare Notification. Specifications and Standards for Food, Food Additives, etc. 1959; No. 370. Available online: https://www.mhlw.go.jp/english/topics/foodsafety/positivelist060228/dl/r01_a.pdf (accessed on 4 September 2020).

14. Taniguchi, A.; Maruyama, R.; Kyogoku, N.; Kimura-Watanabe, H.; Nagao, K.; Kobayashi, R. Preparing method of Tempura batter using millet flour based on basic grain properties. *J. Home Econ. Jpn.* **2018**, *69*, 217–224. (In Japanese)

15. Awatsuhara, R.; Harada, K.; Maeda, T.; Nomura, T.; Nagao, K. Antioxidative activity of the buckwheat polyphenol rutin in combination with ovalbumin. *Mol. Med. Rep.* **2010**, *3*, 121–125. [CrossRef] [PubMed]

16. Japan Oil Chemists' Society. *The JOCS Standard Methods for the Analysis of Fats, Oils and Related Materials. AV*; 2.3.1-2013, AnV; 2.5.3-2013, PC; 2.5.5-2013; Japan Oil Chemists' Society: Tokyo, Japan, 2013.

17. Folin, O.; Denis, W. A colorimetic method for the determination of phenols (and phenol derivatives) in urine. *J. Biol. Chem.* **1912**, *12*, 239–243.

18. Hwang, H.S.; Winkler-Moser, J.K.; Bakota, E.L.; Berhow, M.A.; Liu, S.X. Antioxidant activity of sesamol in soybean oil under frying conditions. *J. Am. Oil Chem. Soc.* **2013**, *90*, 659–666. [CrossRef]

19. Ravi Kiran, C.; Sasidharan, I.; Soban Kumar, D.R.; Sundaresan, A. Influence of natural and synthetic antioxidants on the degradation of Soybean oil at frying temperature. *J. Food Sci. Technol.* **2015**, *52*, 5370–5375. [CrossRef]

20. Srivastava, Y.; Semwal, A.D. A study on monitoring of frying performance and oxidative stability of virgin coconut oil (VCO) during continuous/prolonged deep fat frying process using chemical and FTIR spectroscopy. *J. Food Sci. Technol.* **2015**, *52*, 984–991. [CrossRef]
21. Khatun, M.; Eguchi, S.; Yamaguchi, T.; Takamura, H.; Matoba, T. Effect of thermal treatment on radical-scavenging activity of some spices. *Food Sci. Technol. Res.* **2006**, *12*, 178–185. [CrossRef]
22. Nile, S.H.; Park, S.W. Chromatographic analysis, antioxidant, anti-inflammatory, and xanthine oxidase inhibitory activities of ginger extracts and its reference compounds. *Ind. Crop Prod.* **2015**, *70*, 238–244. [CrossRef]
23. Do, T.T.; Muhlhausler, B.; Box, A.; Able, A.J. Enrichment of antioxidant capacity and vitamin E in pita made from barley. *J. Food Sci.* **2016**, *81*, H777–H785. [CrossRef] [PubMed]
24. Jan, U.; Gani, A.; Ahmad, M.; Shah, U.; Baba, W.N.; Massogi, F.A.; Gani, A.; Wani, I.A.; Wani, S.M. Characterization of cookies made from wheat flour blended with buckwheat flour and effect on antioxidant properties. *J. Food Sci. Technol.* **2015**, *52*, 6334–6344. [CrossRef] [PubMed]
25. Firestone, D. Worldwide Regulation of Frying Fats and Oils. *INFORMS* **1993**, *4*, 1366–1371.
26. Kaunitz, H.; Slanetz, C.A.; Johnson, R.E. Antagonism of fresh fat to the toxicity of heated and aerated cottonseed oil. *J. Nutr.* **1955**, *55*, 577–587. [CrossRef]
27. Idehen, E.; Tang, Y.; Sang, S. Bioactive phytochemicals in barley. *J. Food Drug Anal.* **2017**, *25*, 148–161. [CrossRef]
28. Zhang, G.; Xu, Z.; Gao, Y.; Huang, X.; Zou, Y.; Yang, T. Effects of germination on the nutritional properties, phenolic profiles, and antioxidant activities of buckwheat. *J. Food Sci.* **2015**, *80*, H1111–H1119. [CrossRef]
29. Wang, L.; Chen, J.; Xie, H.; Ju, X.; Liu, R.H. Phytochemical profiles and antioxidant activity of adlay varieties. *J. Agric. Food Chem.* **2013**, *61*, 5103–5113. [CrossRef]

*Article*

# Effect of Sunlight Exposure on Anthocyanin and Non-Anthocyanin Phenolic Levels in Pomegranate Juices by High Resolution Mass Spectrometry Approach

**Vita Di Stefano [1],*, Salvatore Scandurra [2], Antonella Pagliaro [3], Vincenzo Di Martino [2] and Maria Grazia Melilli [2],***

[1]  Department of Biological, Chemical and Pharmaceutical Sciences and Technologies (STEBICEF), University of Palermo, Via Archirafi 32, 90123 Palermo, Italy

[2]  Institute for Agricultural and Forest Systems in the Mediterranean, National Council of Research, Via Empedocle, 58, 95128 Catania, Italy; salvatore.scandurra@cnr.it (S.S.); vincenzo.dimartino@cnr.it (V.D.M.)

[3]  CREA Research Centre for Cereal and Industrial Crops, 95024 Acireale (Catania), Italy; anto.pagliaro90@gmail.com

*  Correspondence: vita.distefano@unipa.it (V.D.S.); mariagrazia.melilli@cnr.it (M.G.M.); Tel.: +39-091-2389-1948 (V.D.S.); +39-095-613-9916 (M.G.M.)

Received: 22 July 2020; Accepted: 20 August 2020; Published: 23 August 2020

**Abstract:** Quali-quantitative analyses of anthocyanins and non-anthocyanin phenolic compounds performed with the use of liquid chromatography coupled with high resolution mass spectrometry, were evaluated in juice of pomegranate fruits ('Dente di Cavallo'), in relation to different light exposures (North, South, West and East). A total of 16 compounds were identified, including phenolic acids, flavonoids, hydrolysable tannins, and anthocyanins, known for their health-promoting effects. Striking differences were observed about the total phenolic content, which was high in juices from fruits with east- and north-facing position, while it was lower in juices facing south. The greatest contents of total flavonoids and anthocyanins were recorded in fruit juices with southern exposure; however, there are no great differences in the content in phenolic acids. Tannins were mainly synthesized in fruit juices with West exposure. The results showed that the position within the tree had no significant effects on color juice, however, it significantly ($p < 0.05$) affected data on fruit weight, soluble sugars and juice yield. Remarkable synergies existed among polyphenols and phytochemicals in pomegranate juice, but collecting fruits with different solar exposure could enhance different health benefits, i.e., the juices with higher polyphenols content could have more anticancer effect or those with higher tannins content could have more antimicrobial effect.

**Keywords:** *Punica granatum*; polyphenols; hydrolysable tannins; flavonoids; Ultra High Performance Liquid Chromatography-Orbitrap-Mass Spectrometry

---

## 1. Introduction

Pomegranate (*Punica granatum* L.) trees are cultivated worldwide in subtropical and tropical regions. Among the largest countries producers are Iran, India, USA, Turkey, Egypt, Italy, Chile, and Spain [1]. The crop is known since ancient time, for its nutritional, medicinal, and ornamental importance [2].

Spain and Italy are the main European producers. In Italy, pomegranate fruits production amounts to around 60,000 tons per year, and the best producers are the Apulia and Sicily regions. The most cultivated varieties are Dente di Cavallo, Mollar, Acco, and Wonderfull [3].

About 50% of the total weight of the pomegranate is made up of peel and carpellar membranes, important sources of active compounds such as flavonoids; elagitannins; proanthocyanidins; polysaccharides; and minerals such as calcium, magnesium, phosphorus and sodium.

The edible component of the fruit represents approximately 50% of the fruit made up of 80% of arils (fleshy part) and 20% of seeds (woody part). The arils contain 85% water; 10% sugars such as glucose and fructose; 1.5% pectin; organic acids such as ascorbic acid, citric acid, and malic acid; and phenolic compounds such as phenolic acids, flavonoids and anthocyanins. The seeds are instead rich in lipids. The oil that can be extracted is 12–20% of the total weight of the seeds and is characterized by a high concentration of polyunsaturated fatty acids such as linoleic acid and linolenic acid. The seeds also contain proteins, vitamins, fiber, pectins, sugars, and phytoestrogens [4].

The market is constantly growing, which is presumably due to the increasing awareness of consumers of the potential health benefits attributed to the pomegranate trees and their processed products [5–7].

Pomegranate juices (PJs) are well known for their beneficial properties, they carry out antioxidant, antimicrobial, anticancer, cholesterol-lowering, anti-atherosclerotic, and anti-diabetic activities [8–16]. As anthocyanin-rich food, pomegranate juices have been used at various concentrations to enhance the color, taste and aroma properties; to increase the health-benefits properties; and to improve shelf lives of the fortified foods [17–19]. The numerous pharmacological studies associated with the regular consumption of PJs in the prevention of certain diseases and the improvement of health conditions, have made it possible to define the fruit as a functional food. Anthocyanins, ellagic acid derivatives, and hydrolysable tannins were detected in pomegranate juices as responsible for antioxidant activity. Anthocyanins are important flower and fruit pigments; they attract pollinators and seed dispersers and protect plant tissues from photo-inhibition and oxidation resulting from photosynthesis [20].

Anthocyanins act as free radical scavengers, thanks to the *o*-diphenol substitution in ring B of anthocyanins (Figure 1) and the conjugated double-bond system, stabilizing radicals due to hydrogen donation. Metal chelation [21] or DNA and protein binding [22] are important in biological systems.

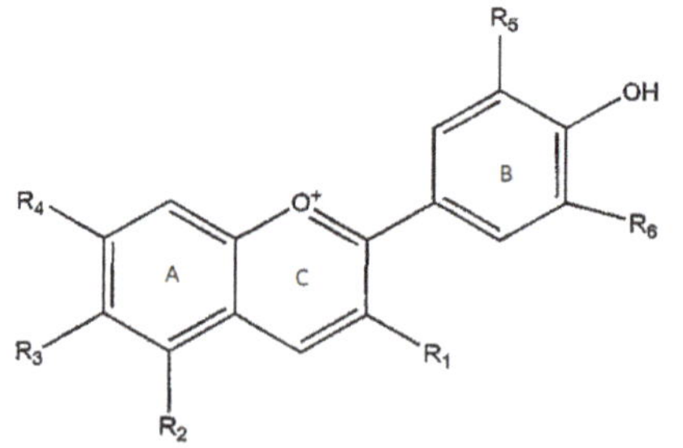

|  | $R_1$ | $R_2$ | $R_3$ | $R_4$ | $R_5$ | $R_6$ |
|---|---|---|---|---|---|---|
| Cyanidin-3-O-glucoside | O-glucose | OH | H | OH | OH | H |
| Delphinidin-3-O-glucoside | O-glucose | OH | H | OH | OH | OH |
| Pelargonidin-3-O-glucoside | O-glucose | OH | H | OH | H | H |
| Cyanidin-3,5-O-diglucoside | O-glucose | O-glucose | H | OH | OH | H |
| Delphinidin-3,5-O-diglucoside | O-glucose | O-glucose | H | OH | OH | OH |
| Pelargonidin-3,5-O-diglucoside | O-glucose | O-glucose | H | OH | H | H |

**Figure 1.** Anthocyanins in pomegranate juices [23].

Delphinidin, cyanidin and pelargonidin glycosides are the most representative anthocyanins in PJs with strong antioxidant activities [23–26].

As is known, plants and their fruits exhibit physiological responses to solar radiation [27,28]. It is believed that exposure to solar radiation stimulates the production of phenolic substances, which absorb UV rays, to protect tissues from DNA-induced damage [29–31].

Direct exposure to sunlight has been shown to regulate the biosynthesis of phenolic compounds by altering the expression of genes that code for enzymes of the biosynthetic pathways [32–35].

On some types of fruit, it has been observed that the content of anthocyanins and quercetin is far superior in the parts exposed to light, while for other important compounds such as catechins, phloridzin and chlorogenic acid, there are no differences between the fruits exposed to light and those exposed to shade [36].

This work aims to enhance the understanding of the relationship between light of solar radiation on the accumulation of total phenolic content, flavonoids and hydrolysable tannins during fruit ripening.

Pomegranate fruits cv "Dente di Cavallo", of an experimental orchard located in south-eastern Sicily, Italy, under natural sun irradiance, with North, South, East, and West exposure (Noto N, Noto S, Noto E and Noto W) were collected and total phenolic contents, total anthocyanin contents, and the levels of hydrolysable tannins in the pomegranate juices were determined.

The ultra-high-performance liquid chromatography and high-resolution quadrupole Orbitrap mass spectrometry (UHPLC–Orbitrap-MS) approach was employed to perform phytochemical analyses in PJs of the fruits growing with different light exposure; moreover, color and physicochemical analysis were carried out.

## 2. Materials and Methods

### 2.1. Plant Material, Harvesting Procedures, Preparation of Pomegranate Juices (PJs) and Instrumental Color Measurement

The experimental analysis was conducted on samples of fruits belonging to the cv "Dente di Cavallo", one of the main cultivars cropped in Sicily. The selected plant material were harvested during November 2018 at the field located in Noto (c.da Fiumara, 36.90° N; 15.05° E, Syracuse, Sicily).

The orchard was established in 1993; hence, trees are 25 years old, it is 2.5 ha. Pomegranate trees were planted at a spacing of 4 m × 4 m. They are drip irrigated, and standard cultural practices are performed (pruning, thinning, fertilization and pest control treatments). Ten trees have been sampled and for each tree, five fruits per four geographical orientation of tree (North-N, South-S, East-E and West-W respectively). Each tree is a biological repetition.

The harvest was performed according the ripening stage, about 180 days after full bloom. Samples were taken from 12:00 h to 14:30 h.

After picking, fruits were immediately transported to the laboratory. Each fruit was carefully cut at the equatorial zone with a sharpened knife, and then intact arils were manually obtained from whole fruits, and the juices were obtained by squeezing them by mechanical press and stored at −20 °C. Weight of the fruits (1 mg accuracy on a balance), weight of the arils per fruit, juice yield (%) and the color of juices (L *, a *, b *) by colorimeter Minolta CR 400 (Konica Minolta, Milan, Italy) were determined [23,37].

### 2.2. Total Phenolic Content (TPC)

Phenolic contents of PJs were determined by Folin–Ciocalteu's method, and total polyphenols content (TPC) was expressed as grams gallic acid equivalents (GAE) per $L^{-1}$ of juice [23].

### 2.3. Determination of Phenolic Compounds by UHPLC–Orbitrap-MS

Identification of polyphenols (phenolic acids, flavonoids, tannins and anthocyanin-derived pigments) in PJs was based on a reported procedure and the UHPLC–Orbitrap-MS method previously developed [23].

The MS detection was conducted in two acquisition modes: full scan (positive and negative ion modes) and targeted selected ion monitoring. For targeted selected ion monitoring analyses, a mass inclusion list containing exact masses and expected retention times of target phenolic acids, flavonoids, tannins, and anthocyanins analytes was built and applied (Supplementary materials Figures S1 and S3). Phenolic compounds were also identified by MS/MS data-dependent dd-ms$^2$ scanning mode.

*2.4. Statistical Analysis*

Data were submitted to the Bartlett's test and then analyzed using analysis of variance (ANOVA). Means were statistically separated on the basis of Student-Newmann-Kewls test.

## 3. Results

Single fruit weight resulted in being 297 ± 10.97 g, with variations due to solar radiation, influencing the weight of arils and the juice yield (Table 1). Soluble sugars resulted on average 13.4 ± 0.4 °Brix, picked in 14.8 ± 0.77 °Brix in fruit collected in W and S exposure. The color of juice did not show statistical differences in relation to the solar radiation.

**Table 1.** Fruit weight, juice yield, soluble sugars, and juice color coordinates.

| | Fruit Weight (g) | Arils Weight (g Fruit$^{-1}$) | Juice Yield (%) | Soluble Sugar (°Brix) | L * | a * | b * |
|---|---|---|---|---|---|---|---|
| Noto E | 276 ± 13.25 [b] | 170 ± 2.04 [b] | 56.9 ± 1.19 [b] | 13.2 ± 0.16 [b] | 40.7 ± 2.12 | 11.5 ± 0.37 | 21.88 ± 0.70 |
| Noto N | 319 ± 13.40 [a] | 189 ± 4.54 [ab] | 61.5 ± 2.58 [a] | 13.7 ± 0.33 [b] | 40.6 ± 0.97 | 11.4 ± 0.48 | 21.75 ± 0.83 |
| Noto W | 282 ± 9.02 [b] | 162 ± 8.26 [b] | 63.2 ± 1.52 [a] | 14.7 ± 0.35 [a] | 41.2 ± 1.77 | 11.7 ± 0.47 | 21.85 ± 0.59 |
| Noto S | 342 ± 8.21 [a] | 218 ± 6.10 [a] | 66.7 ± 3.47 [a] | 14.8 ± 0.77 [a] | 41.3 ±1.16 | 11.9 ± 0.44 | 21.7 ± 1.24 |
| Means | 297 ± 10.97 | 179 ± 5.24 | 59.2 ± 2.19 | 13.4 ± 0.40 | 40.9 ± 1.50 | 11.6 ± 0.44 | 21.8 ± 0.84 |

Different letters within the same column indicate differences at $p < 0.05$. All values are mean ± SD of five independent measurements of each sample.

Total phenolic content (TPC) is an important quality parameter of PJs above all for the organoleptic characteristics, especially color and taste properties and, of course, for the high antioxidant activity. Light exposure during growth greatly affect phenolics accumulation. Phenolic compounds are important secondary metabolites with antioxidant capacity in fruit and were found to be associated with resistance to scald development in fruits [38]. The Folin–Ciocalteu test indicated a high TPC in all samples of juice analyzed. TPC levels ranged from 4.82 g GAE L$^{-1}$ to 11.06 g GAE L$^{-1}$, with the highest amount in Noto N (11.06 ± 0.37 g GAE L$^{-1}$), followed by Noto E (11.02 ± 0.10 g GAE L$^{-1}$) (Figure 2).

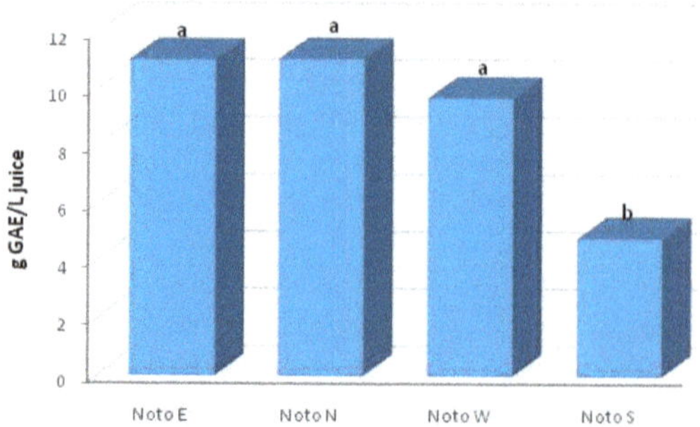


**Figure 2.** TPC (g GAE L$^{-1}$ juice) of the samples. Different letters indicate differences at $p < 0.05$; $n = 5$.

Elagitannins, such as punicalagins and punicalins, are the main hydrolysable tannins found in PJs, while, among anthocyanins, cyanidin, pelargonidin, and delphinidin, glucosides are the main bio-molecules characterizing the PJs. The factor "position within the tree" has been studied, besides the differences in TPC, also for the evaluation of the profile in polyphenols in juices.

Identification of polyphenols phenolic acids, flavonoids, and hydrolysable tannins in PJs have been built on reported UHPLC–Orbitrap-MS approach [23].

The presented data in Table 2 show the average values of the fruit juices in the four positions studied (East, West, South and North exposure), for anthocyanins content.

The anthocyanin profile was determined for PJs of different samples, and the chemical structures were determined by the mass spectra in positive ionization mode at different fragmentation energy, and by comparison with literature data.

An inclusion list (Supplementary Materials, Figure S1), in which were reported the molecular formula together with the accurate mass of anthocyanins known, was prepared [23].

Six anthocyanins with a retention time of between 12.20 min and 14.06 min were identified in all the PJs samples. The two available standards were cyanidin-3-O-glucoside and cyanidin-3,5-O-diglucoside. Their use was useful to recognize their presence in the juices analyzed by comparing the retention times. Furthermore, the exact mass and the study of their MS/MS fragmentation was crucial for the identification of the glycosylated anthocyanins (Supplementary Materials, Figure S2).

The quantitative analysis of anthocyanins performed with the use of liquid chromatography coupled with high resolution mass spectrometry, combines the quadrupole selection of the precursor ion with the Orbitrap detection at high resolution and mass precision.

The results proved that the juices studied had the same phenolic qualitative profile but their relative abundances were considerably different. For their quantification, cyanidin 3-O-glucoside $m/z$ 449.1078 $[M]^+$ was the reference compound used for quantification of pelargonidin-3-O-glucoside, delphinidin-3-O-glucoside (Table 3). Cyanidin 3,5-O-diglucoside $m/z$ 611.1606 $[M]^+$ is the reference compound in positive mode for pelargonidin-3,5-O-diglucoside and delphinidin-3,5-O-diglucoside (Supplementary Materials, Table S1).

**Table 2.** Anthocyanins content mg $L^{-1}$ of juice of different fruits (SD).

| Phenolic Compounds | Noto E | Noto N | Noto W | Noto S |
|---|---|---|---|---|
| Cyanidin-3-O-glucoside * | 69.85 ± 1.06 [cd] | 145.81 ± 1.77 [b] | 87.47 ± 0.39 [d] | 169.74 ± 0.25 [a] |
| Delphinidin-3-O-glucoside | 21.08 ± 0.19 [cd] | 41.93 ± 0.19 [b] | 19.66 ± 0.23 [d] | 45.83 ± 0.06 [a] |
| Pelargonidin-3-O-glucoside | 8.61 ± 0.08 [d] | 17.41 ± 0.10 [b] | 9.91 ± 0.03 [c] | 21.66 ± 0.03 [a] |
| Pelargonidin-3,5-O-diglucoside | 36.87 ± 0.22 [c] | 50.21 ± 0.17 [b] | 24.52 ± 0.02 [d] | 56.02 ± 0.11 [a] |
| Cyanidin-3,5-O-diglucoside * | 374.00 ± 1.97 [c] | 639.78 ± 1.02 [b] | 305.63 ± 0.14 [d] | 765.69 ± 0.51 [a] |
| Delphinidin-3,5-O-diglucoside | 118.66 ± 1.42 [c] | 212.11 ± 0.21 [b] | 69.43 ± 0.27 [d] | 225.98 ± 0.14 [a] |
| Total anthocyanins | 510.41 | 1107.25 | 516.62 | 1284.92 |

Different letters in row indicate differences at $p < 0.05$. * Commercial standard used for their quantification.

In particular, Noto S shows the highest content of total anthocyanins (Table 3), followed by Noto N, Noto W and Noto E. Delphinidin, cyanidin and pelargonidin 3-glucosides and 3,5-diglucosides, which are responsible for the red–purple colour of juices, represent the main anthocyanins of the PJs. In decreasing order, the concentrations of anthocyanins in juices were: cyanidin-3,5-O-diglucoside in a range of 305.63–765.69 mg $L^{-1}$; delphinidin-3,5-O diglucoside, 69.43–225.98 mg $L^{-1}$; pelargonidin-3,5-O-diglucoside, 24.52–56.02 mg $L^{-1}$; cyanidin-3-O-glucoside, 69.85–169.74 mg $L^{-1}$; pelargonidin-3-O-glucoside, 8.61–21.66 mg $L^{-1}$; and delphinidin-3-O-diglucoside, 19.66–45.83 mg $L^{-1}$.

In the current study, the levels of anthocyanins obtained are greater than those reported by literature for other cultivar grown in Sicily [39]. In addition, great differences were observed in function of sunlight exposure; Noto S showed higher anthocyanins content (1284.92 mg $L^{-1}$) followed by Noto N (1107.25 mg $L^{-1}$).

An inclusion list was prepared before UHPLC-MS analysis in negative mode of Noto S, Noto N, Noto E, and Noto W samples (Supplementary Materials, Figure S3).

**Table 3.** Identified anthocyanins in positive ionization in pomegranate juices by UHPLC-Orbitrap-MS approach.

| Compounds | $[M]^+$ *m/z* | MS/MS *m/z* | Molecular Formula | Retention Time (min) |
|---|---|---|---|---|
| Pelargonidin-3-O-glucoside | 433.1129 | 271.0593 | $C_{21}H_{21}O_{10}$ | 14.08 |
| Cyanidin-3-O-glucoside * | 449.1078 | 287.0542 | $C_{21}H_{21}O_{11}$ | 13.56 |
| Delphinidin-3-O-glucoside | 465.1028 | 303.0492 | $C_{21}H_{21}O_{12}$ | 13.17 |
| Pelargonidin-3,5-O-diglucoside | 595.1658 | 433.1120271.0594 | $C_{27}H_{31}O_{15}$ | 13.44 |
| Cyanidin-3,5-O-diglucoside * | 611.1606 | 449.1018287.0542 | $C_{27}H_{31}O_{16}$ | 12.79 |
| Delphinidin-3,5-O-diglucoside | 627.1556 | 465.1018303.0492 | $C_{27}H_{31}O_{17}$ | 12.20 |

MS/MS = Fragmented anthocyanidin molecular weight. This fragment ion is cleavage product free from the sugar moiety. * Commercial standard used for their quantification.

Mass spectra in Full MS, accurate mass and the spectra in MS/MS made it possible to structurally recognize phenolic compounds with retention times between 8.52 min and 20.67 min. Ten compounds have been identified belonging to the class of phenolic acids, flavonoids and tannins, already known in the literature for pomegranate [40].

Figure S4 shows the retention times of the compounds with exact mass collected in the inclusion list. Table 4 shows the tannins, phenolic acids and flavonoids identified in PJ samples in negative mode.

**Table 4.** Phenolic compounds identified in negative ionization in PJs by UHPLC-Orbitrap-MS approach.

| Compounds | $[M-H]^-$ *m/z* | MS/MS *m/z* | Molecular Formula | Retention Time (min) |
|---|---|---|---|---|
| Gallic acid * | 169.01344 | 125.02334 | $C_7H_6O_5$ | 11.12 |
| Kaempferol-3-O-glucoside | 447.09414 | 285.00387 | $C_{21}H_{20}O_{11}$ | 13.56 |
| Quercetin 3-O-hexoside | 463.08932 | 300.9990 | $C_{21}H_{20}O_{12}$ | 13.16 |
| Rutin * | 609.14805 | 447.0571284.9679 | $C_{27}H_{30}O_{16}$ | 12.87 |
| Vanillic acid hexoside | 329.08895 | 167.03417101.02327 | $C_{14}H_{18}O_9$ | 13.09 |
| Ferulic acid hexoside | 355.10476 | 175.03937160.0130 | $C_{16}H_{20}O_9$ | 14.25 |
| Ellagic acid pentoside | 433.04277 | 300.9990 | $C_{19}H_{14}O_{12}$ | 19.85 |
| Ellagic acid deoxyhexoside | 447.05824 | 300.9990 | $C_{20}H_{16}O_{12}$ | 20.67 |
| Corilagin | 633.07469 | 470.98416 | $C_{27}H_{22}O_{18}$ | 14.20 |
| Lagerstannin C | 649.07055 | 486.97901 | $C_{27}H_{22}O_{19}$ | 8.52 |

* Commercial reference compounds used for quantification of individual phenolic constituents.

Rutin, gallic acid and cyanidin 3-O-glucoside *m/z* 447.093 $[M-2H]^-$ were used as reference standards. Cyanidin 3-O-glucoside in negative mode was useful for quantification of vanillic acid hexoside, ferulic acid hexoside, ellagic acid pentoside, ellagic acid deoxyhexoside, kaempferol 3-O glucoside, quercetin 3-O-hexoside, corilagin, and lagerstannin C.

Each phenolic compound was quantified using its respective reference compound of the calibration curves [23]; linear equations and linear regression coefficient of the standards used are reported in the Supplementary Materials (Table S1).

Table 5 shows the quantities of total and individual phenolic compounds expressed as mg $L^{-1}$ contained in the different pomegranate juices.

**Table 5.** Phenolic content expressed as mg L$^{-1}$ of PJs.

| Phenolic Compounds | Noto E | Noto N | Noto W | Noto S |
|---|---|---|---|---|
| Gallic acid * | 1.12 ± 0.00 [b] | 0.9 ± 0.04 [b] | 1.21 ± 0.03 [b] | 0.94 ± 0.05 [b] |
| Kaempferol-3-O-glucoside | 74.94 ± 0.09 [c] | 103.59 ± 0.19 [b] | 62.99 ± 0.11 [d] | 130.80 ± 0.16 [a] |
| Quercetin 3-O-hexoside | 26.60 ± 0.06 [c] | 33.60 ± 0.03 [b] | 16.96 ± 0.02 [ef] | 37.55 ± 0.04 [a] |
| Rutin * | 0.74 ± 0.02 [b] | 1.02 ± 0.00 [a] | 0.77 ± 0.00 [b] | 1.13 ± 0.01 [a] |
| Vanillic acid hexoside | 67.75 ± 0.12 [b] | 70.34 ± 0.01 [a] | 47.52 ± 0.07 [e] | 58.00 ± 0.01 [c] |
| Ferulic acid hexoside | 60.35 ± 0.01 [a] | 59.97 ± 0.04 [a] | 33.93 ± 0.10 [c] | 53.77 ± 0.26 [b] |
| Ellagic acid pentoside | 10.48 ± 0.26 [b] | 10.60 ± 0.04 [b] | 13.18 ± 0.08 [a] | 12.36 ± 0.08 [a] |
| Ellagic acid deossihexoside | 26.64 ± 0.09 [c] | 23.77 ± 0.03 [d] | 29.94 ± 0.06 [a] | 28.32 ± 0.04 [b] |
| Corilagin | 8.70 ± 0.17 [b] | 6.86 ± 0.02 [c] | 12.66 ± 0.03 [a] | 7.90 ± 0.02 [bc] |
| Lagerstannin C | 10.03 ± 0.05 [d] | 12.66 ± 0.04 [c] | 22.01 ± 0.06 [a] | 13.01 ± 0.06 [c] |
| Total phenolic acids | 1.12 | 0.90 | 1.21 | 0.94 |
| Total phenolic acids hexoside | 128.10 | 130.31 | 81.45 | 111.77 |
| Total flavonoids | 102.28 | 138.21 | 80.72 | 169.48 |
| Total hydrolysable tannins | 55.85 | 53.89 | 77.79 | 61.59 |

Different letters in row indicate differences at $p < 0.05$. * Commercial reference compounds used for quantification of individual phenolic constituents.

The compounds are listed below according to their chemical class, and the concentrations are expressed in mg L$^{-1}$ of juice. Among the hydrolysable tannins, similar levels of lagerstannin C and corilagin (10.03–22.01 mg L$^{-1}$ and 6.86–2.66 mg L$^{-1}$, respectively) were found in all samples. Noto W sample has shown a higher content of ellagic acids derivative (29.94 mg L$^{-1}$) and overall a higher concentration of tannins.

The highest concentration of flavonoids was found in the juices of the Noto N and Noto S samples (138.21 and 169.48 mg L$^{-1}$ respectively). Noto N and Noto E had a higher content of phenolic acids hexoside (130.31 and 128.10 mg L$^{-1}$, respectively).

In the present study, pomegranate fruits with South and North exposure had juices with significantly higher concentrations of anthocyanins, Noto S (1284.92 mg L$^{-1}$), followed by Noto N (1107.25 mg L$^{-1}$). Noto S showed the higher content in flavonoids (169.48 mg L$^{-1}$). Noto N with 130.31 mg L$^{-1}$ shower the highest concentration of total phenolic acids hexoside. The relative distribution of each class of phenolic compounds is reported in Table 3. Noto E synthesized mainly total hydrolysable tannins followed by Noto N. Noto S accumulated mainly total flavonoids (Figure 3).

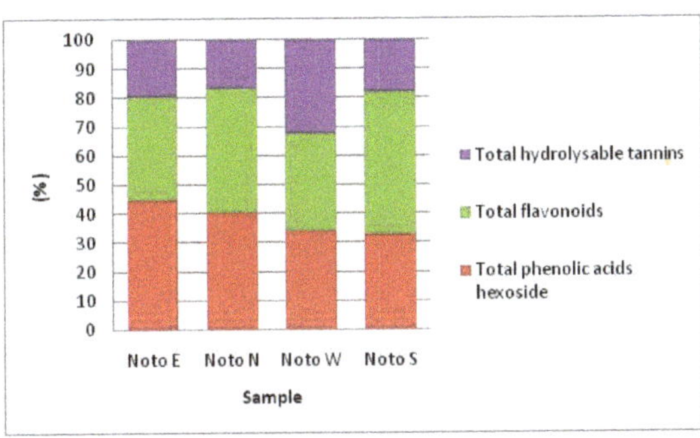


**Figure 3.** Partitioning (% of total) of the main phenolic compounds in pomegranate juices in relation to solar exposure.

## 4. Discussion

The polyphenol composition of PJs has been characterized as being complex and unique, comprising mainly anthocyanins, phenolic acids and hydrolysable tannins. In this work, we used UHPLC-Orbitrap-MS method for the analysis of both. The main hydrolysable tannins, flavonoids, phenolic acids and anthocyanins in juices of fruits with different sunlight exposures, in order to provide a simple and easily applicable method for quality control of the juices, were identified. Juices of all the samples present similar HPLC-MS profiles, with great difference mainly regarding some anthocyanins contents. The values of TPC recorded in PJs are greater than those reported in Sicilian pomegranate juices by Todaro et al. [39], which ranged from 0.95 to 3.10 g GAE L$^{-1}$, but are similar to the other data reported in literature [41,42]. Differences in reported total phenolic compounds content could be partially explained by pomegranate variety, different water content, environmental growing conditions or ripening stage [43], and pre- and post-harvest treatments [44].

Phenolic compounds are important secondary metabolites with antioxidant capacity associated with resistance to the development of sunburn on the fruit surface [45]. It has been reported that the activity of enzymes involved in the phenolic metabolism pathway can be photonically upregulated [46]. The photoprotective function of anthocyanins has been well documented [47].

Great differences were observed in function of sunlight exposure; PJ with South exposure showed higher anthocyanins content followed by North exposure fruit juices. The highest concentration of flavonoids was found in the juices of the Noto S sample, with a relative percentage of 49.3% of total phenolic compounds. Noto N showed the highest content of phenolic acids hexoside (44.6%), while fruit exposed to West showed 32% in tannins against values lower than 18% in the other samples. Fruit exposure to sun radiation in S, E and W positions might reduce the concentration of flavonoids and phenolic compounds due to elevated temperature, which causes a reduction in their synthesis or even partial degradation. Anthocyanins are typical potent antioxidants and may contribute to hydrogen peroxide scavenging during direct light sun exposure [46]. Based on the markedly higher anthocyanin contents in sun-exposed pomegranate fruits (Noto S) than in the shaded one, we suggest that they might be the main phenolic compounds responsible for protecting fruit from UV/high light damage, where as other class of phenolics are involved to a lesser extent in photoprotection of fruits.

## 5. Conclusions

The obtained results provided evidence that relevant differences on the patterns of anthocyanins in pomegranate fruits can be explained by the effect of light exposure.

Enhanced knowledge of the role played by sunlight intensity on the accumulation of phenolic compounds enables support of cultural practices for a given phenolic profile in pomegranate fruits and fosters the production of high-quality juices.

This study gave new information about the different compositions in bio-molecules with health benefit of PJs in relation to solar exposure of the pomegranate fruits.

**Supplementary Materials:** The following are available online at http://www.mdpi.com/2304-8158/9/9/1161/s1, Figure S1: Inclusion list with molecular formula and accurate mass of anthocyanins in positive ionization, by UHPLC–Orbitrap-MS; Figure S2: Retention time (min) of anthocyanins identified in PJs; Figure S3: Inclusion list of phenolic acids, flavonoids and tannins with molecular formula and accurate mass of anthocyanins in negative ionization, by UHPLC–Orbitrap-MS; Figure S4: Retention time (min) of phenolic acids, flavonoids and tannins identified in PJs; Table S1: *m/z*, molecular formula, linear regression model and coefficient of determination of external standards used for calibration.

**Author Contributions:** Conceptualization, V.D.S. and M.G.M.; methodology, V.D.S.; validation, V.D.S. and M.G.M.; formal analysis, A.P., S.S. and M.G.M.; investigation, V.D.S.; resources, V.D.M.; data curation, V.D. and M.G.M.; writing—original draft preparation, V.D.S. and M.G.M.; writing—review and editing, V.D.S. and M.G.M.; supervision, V.D.S. and M.G.M.; project administration, M.G.M.; funding acquisition, M.G.M. All authors have read and agreed to the published version of the manuscript.

**Funding:** This work has been supported by National Research Council, CNR—DISBA project NutrAge (project nr. 7022).

**Acknowledgments:** The authors thank the Mass Spectrometry laboratory of Advanced Technologies Network (ATeN) Center at the University of Palermo for technical support. Thanks to Dott. Rosa Pitonzo for her valuable and constructive suggestions during the development of this research work. The authors also thank Dott. Lucia Sollima for her administrative support.

**Conflicts of Interest:** The authors declare no conflict of interest.

## References


1. Stover, E.; Mercure, E.W. The pomegranate: A new look at the fruit of paradise. *HortScience* **2007**, *42*, 1088–1092. [CrossRef]
2. Qin, G.; Liu, C.; Li, J.; Qi, Y.; Gao, Z.; Zhang, X.; Yi, X.; Pan, H.; Ming, R.; Xu, Y. Diversity of metabolite accumulation patterns in inner and outer seed coats of pomegranate: Exploring their relationship with genetic mechanisms of seed coat development. *Hortic. Res.* **2020**, *7*. [CrossRef] [PubMed]
3. Cossi, F. Melograno, potenzialità e limiti di un antico frutto italiano. *Riv. Fruttic. Ortofloric.* **2017**, *81*, 52–63.
4. Viuda-Martos, M.; Fernández-Lóaez, J.; Pérez-álvarez, J.A. Pomegranate and its Many Functional Components as Related to Human Health: A Review. *Compr. Rev. Food Sci. Food Saf.* **2010**, *9*, 635–654. [CrossRef]
5. Bartual, J.; Fernandez-Zamudio, M.A.; De Miguel, M.D. Situation of the production, research and economics of the pomegranate industry in Spain. *Acta Hortic.* **2015**, *1089*, 345–349. [CrossRef]
6. Rymon, D. Mapping features of the global pomegranate market. *Acta Hortic.* **2011**, 599–601. [CrossRef]
7. Işık Özgüven, A.; Gültekin, U.; Gözlekçi, S.; Yilmaz, I.; Yilmaz, C.; Küçük, E.; Imrak, B.; Korkmaz, C. A review of the economics and the marketing of the pomegranate industry in Turkey. *Acta Hortic.* **2015**, 221–228. [CrossRef]
8. Katz, S.R.; Newman, R.A.; Lansky, E.P. Punica granatum: Heuristic Treatment for Diabetes Mellitus. *J. Med. Food* **2007**, *10*, 213–217. [CrossRef]
9. Esmaillzadeh, A.; Tahbaz, F.; Gaieni, I.; Alavi-Majd, H.; Azadbakht, L. Cholesterol-Lowering Effect of Concentrated Pomegranate Juice Consumption in Type II Diabetic Patients with Hyperlipidemia. *Int. J. Vitam. Nutr. Res.* **2006**, *76*, 147–151. [CrossRef]
10. Aviram, M.; Dornfeld, L.; Rosenblat, M.; Volkova, N.; Kaplan, M.; Coleman, R.; Hayek, T.; Presser, D.; Fuhrman, B. Pomegranate juice consumption reduces oxidative stress, atherogenic modifications to LDL, and platelet aggregation: Studies in humans and in atherosclerotic apolipoprotein E-deficient mice. *Am. J. Clin. Nutr.* **2000**, *71*, 1062–1076. [CrossRef]
11. Davidson, M.H.; Maki, K.C.; Dicklin, M.R.; Feinstein, S.B.; Witchger, M.S.; Bell, M.; McGuire, D.K.; Provost, J.C.; Liker, H.; Aviram, M. Effects of Consumption of Pomegranate Juice on Carotid Intima-Media Thickness in Men and Women at Moderate Risk for Coronary Heart Disease. *Am. J. Cardiol.* **2009**, *104*, 936–942. [CrossRef] [PubMed]
12. Aviram, M.; Rosenblat, M.; Gaitini, D.; Nitecki, S.; Hoffman, A.; Dornfeld, L.; Volkova, N.; Presser, D.; Attias, J.; Liker, H.; et al. Pomegranate juice consumption for 3 years by patients with carotid artery stenosis reduces common carotid intima-media thickness, blood pressure and LDL oxidation. *Clin. Nutr.* **2004**, *23*, 423–433. [CrossRef] [PubMed]
13. Lei, F.; Zhang, X.N.; Wang, W.; Xing, D.M.; Xie, W.D.; Su, H.; Du, L.J. Evidence of anti-obesity effects of the pomegranate leaf extract in high-fat diet induced obese mice. *Int. J. Obes.* **2007**, *31*, 1023–1029. [CrossRef] [PubMed]
14. Boussetta, T.; Raad, H.; Lettéron, P.; Gougerot-Pocidalo, M.A.; Marie, J.C.; Driss, F.; El-Benna, J. Punicic acid a conjugated linolenic acid inhibits TNFα-induced neutrophil hyperactivation and protects from experimental colon inflammation in rats. *PLoS ONE* **2009**, *4*. [CrossRef] [PubMed]
15. Romier-Crouzet, B.; Van De Walle, J.; During, A.; Joly, A.; Rousseau, C.; Henry, O.; Larondelle, Y.; Schneider, Y.J. Inhibition of inflammatory mediators by polyphenolic plant extracts in human intestinal Caco-2 cells. *Food Chem. Toxicol.* **2009**, *47*, 1221–1230. [CrossRef] [PubMed]
16. Reddy, M.K.; Gupta, S.K.; Jacob, M.R.; Khan, S.I.; Ferreira, D. Antioxidant, antimalarial and antimicrobial activities of tannin-rich fractions, ellagitannins and phenolic acids from *Punica granatum* L. *Planta Med.* **2007**, *73*, 461–467. [CrossRef]
17. Gerardi, C.; Albano, C.; Calabriso, N.; Carluccio, M.A.; Durante, M.; Mita, G.; Renna, M.; Serio, F.; Blando, F. Techno-functional properties of tomato puree fortified with anthocyanin pigments. *Food Chem.* **2018**, *240*, 1184–1192. [CrossRef]
18. Kabakcı, S.A.; Türkyılmaz, M.; Özkan, M. Changes in the quality of kefir fortified with anthocyanin-rich juices during storage. *Food Chem.* **2020**, *326*, 126977. [CrossRef]
19. El-Seideek, L.; Zaied, S.F.; Hassan, M.I.; Elgammal, M.H. Antimicrobial, biochemical, organoleptic and stability properties of cookies fortified by pomegranate juice during storage. *Res. J. Pharm. Biol. Chem. Sci.* **2016**, *7*, 288–299.

20. Lee, D.W.; Gould, K.S. *Anthocyanins in Leaves and Other Vegetative Organs: An introduction*; Elsevier B.V.: Amsterdam, The Netherland, 2002; pp. 1–16. ISBN 9780080953052.

21. Kühnau, J. The flavonoids. A class of semi-essential food components: Their role in human nutrition. *World Rev. Nutr. Diet.* **1976**, *24*, 117–191.

22. Jaldappagari, S.; Motohashi, N.; Gangeenahalli, M.P.; Naismith, J.H. Bioactive Mechanism of Interaction Between Anthocyanins and Macromolecules Like DNA and Proteins. In *Bioactive Heterocycles VI.*; Springer: Berlin/Heidelberg, Germany, 2008; pp. 49–65.

23. Di Stefano, V.; Pitonzo, R.; Novara, M.E.; Bongiorno, D.; Indelicato, S.; Gentile, C.; Avellone, G.; Bognanni, R.; Scandurra, S.; Melilli, M.G. Antioxidant activity and phenolic composition in pomegranate (*Punica granatum* L.) genotypes from south Italy by UHPLC-Orbitrap-MS approach. *J. Sci. Food Agric.* **2019**, *99*, 1038–1045. [CrossRef] [PubMed]

24. Fernandes, L.; Pereira, J.A.; Lopéz-Cortés, I.; Salazar, D.M.; González-Álvarez, J.; Ramalhosa, E. Physicochemical composition and antioxidant activity of several pomegranate (*Punica granatum* L.) cultivars grown in Spain. *Eur. Food Res. Technol.* **2017**, *243*, 1799–1814. [CrossRef]

25. Fischer, U.A.; Dettmann, J.S.; Carle, R.; Kammerer, D.R. Impact of processing and storage on the phenolic profiles and contents of pomegranate (*Punica granatum* L.) juices. *Eur. Food Res. Technol.* **2011**, *233*, 797–816. [CrossRef]

26. Gómez-Caravaca, A.M.; Verardo, V.; Toselli, M.; Segura-Carretero, A.; Fernández-Gutiérrez, A.; Caboni, M.F. Determination of the Major Phenolic Compounds in Pomegranate Juices by HPLC–DAD–ESI-MS. *J. Agric. Food Chem.* **2013**, *61*, 5328–5337. [CrossRef] [PubMed]

27. Ballaré, C.L. Light Regulation of Plant Defense. *Annu. Rev. Plant Biol.* **2014**, *65*, 335–363. [CrossRef] [PubMed]

28. Zoratti, L.; Karppinen, K.; Escobar, A.L.; Häggman, H.; Jaakola, L. Light-controlled flavonoid biosynthesis in fruits. *Front. Plant Sci.* **2014**, *5*, 534. [CrossRef]

29. Li, J.-H.; Guan, L.; Fan, P.-G.; Li, S.-H.; Wu, B.-H. Effect of Sunlight Exclusion at Different Phenological Stages on Anthocyanin Accumulation in Red Grape Clusters. *Am. J. Enol. Vitic.* **2013**, *64*, 349–356. [CrossRef]

30. Winkel-Shirley, B. Biosynthesis of flavonoids and effects of stress. *Curr. Opin. Plant Biol.* **2002**, *5*, 218–223. [CrossRef]

31. Zlatev, Z.S.; Lidon, F.J.C.; Kaimakanova, M. Plant physiological responses to UV-B radiation. *Emirates J. Food Agric.* **2012**, *24*, 481–501. [CrossRef]

32. Azuma, A.; Yakushiji, H.; Koshita, Y.; Kobayashi, S. Flavonoid biosynthesis-related genes in grape skin are differentially regulated by temperature and light conditions. *Planta* **2012**, *236*, 1067–1080. [CrossRef]

33. Matus, J.T.; Loyola, R.; Vega, A.; Peña-Neira, A.; Bordeu, E.; Arce-Johnson, P.; Alcalde, J.A. Post-veraison sunlight exposure induces MYB-mediated transcriptional regulation of anthocyanin and flavonol synthesis in berry skins of Vitis vinifera. *J. Exp. Bot.* **2009**, *60*, 853–867. [CrossRef] [PubMed]

34. Zhang, Z.Z.; Li, X.X.; Chu, Y.N.; Zhang, M.X.; Wen, Y.Q.; Duan, C.Q.; Pan, Q.H. Three types of ultraviolet irradiation differentially promote expression of shikimate pathway genes and production of anthocyanins in grape berries. *Plant Physiol. Biochem.* **2012**, *57*, 74–83. [CrossRef]

35. Martínez-Lüscher, J.; Sánchez-Díaz, M.; Delrot, S.; Aguirreolea, J.; Pascual, I.; Gomès, E. Ultraviolet-B radiation and water deficit interact to alter flavonol and anthocyanin profiles in grapevine berries through transcriptomic regulation. *Plant Cell Physiol.* **2014**, *55*, 1925–1936. [CrossRef] [PubMed]

36. Singleton, V.L.; Rossi, J.A.J. Colorimetry to total phenolics with phosphomolybdic acid reagents. *Am. J. Enol. Vinic.* **1965**, *16*, 144–158.

37. Melilli, M.G.; Di Stefano, V.; Sciacca, F.; Pagliaro, A.; Bognanni, R.; Scandurra, S.; Virzì, N.; Gentile, C.; Palumbo, M. Improvement of Fatty Acid Profile in Durum Wheat Breads Supplemented with *Portulaca oleracea* L. Quality Traits of Purslane-Fortified Bread. *Foods* **2020**, *9*, 764. [CrossRef]

38. Barden, C.L.; Bramlage, W.J. Accumulation of antioxidants in apple peel as related to preharvest factors and superficial scald susceptibility of the fruit. *J. Am. Soc. Hortic. Sci.* **1994**, *119*, 264–269. [CrossRef]

39. Todaro, A.; Cavallaro, R.; Lamalfa, S.; Continella, A.; Gentile, A.; Fischer, U.A.; Carle, R.; Spagna, G. Anthocyanin profile and antioxidant activity of freshly squeezed pomegranate (*Punica Granatum* L.) Juices of sicilian and Spanish provenances. *Ital. J. Food Sci.* **2016**, *28*, 464–479. [CrossRef]

40. Awad, M.A.; De Jager, A.; Van Westing, L.M. Flavonoid and chlorogenic acid levels in apple fruit: Characterisation of variation. *Sci. Hortic.* **2000**, *83*, 249–263. [CrossRef]

41. Tezcan, F.; Gültekin-Özgüven, M.; Diken, T.; Özçelik, B.; Erim, F.B. Antioxidant activity and total phenolic, organic acid and sugar content in commercial pomegranate juices. *Food Chem.* **2009**, *115*, 873–877. [CrossRef]
42. Kalaycıoğlu, Z.; Erim, F.B. Total phenolic contents, antioxidant activities, and bioactive ingredients of juices from pomegranate cultivars worldwide. *Food Chem.* **2017**, *221*, 496–507. [CrossRef]
43. Munera, S.; Hernández, F.; Aleixos, N.; Cubero, S.; Blasco, J. Maturity monitoring of intact fruit and arils of pomegranate cv. 'Mollar de Elche' using machine vision and chemometrics. *Postharvest Biol. Technol.* **2019**, *156*, 110936. [CrossRef]
44. García-Pastor, M.E.; Serrano, M.; Guillén, F.; Zapata, P.J.; Valero, D. Preharvest or a combination of preharvest and postharvest treatments with methyl jasmonate reduced chilling injury, by maintaining higher unsaturated fatty acids, and increased aril colour and phenolics content in pomegranate. *Postharvest Biol. Technol.* **2020**, *167*, 111226. [CrossRef]
45. Barden, C.L.; Bramlage, W.J. Separating the effects of low temperature, ripening, and light on loss of scald susceptibility in apples before harvest. *J. Am. Soc. Hortic. Sci.* **1994**, *119*, 54–58. [CrossRef]
46. Li, P.; Ma, F.; Cheng, L. Primary and secondary metabolism in the sun-exposed peel and the shaded peel of apple fruit. *Physiol. Plant.* **2013**, *148*, 9–24. [CrossRef] [PubMed]
47. Ju, Z.; Yuan, Y.; Liu, C.; Zhan, S.; Wang, M. Relationships among simple phenol, flavonoid and anthocyanin in apple fruit peel at harvest and scald susceptibility. *Postharvest Biol. Technol.* **1996**, *8*, 83–93. [CrossRef]

*Article*

# The Bioactive Potential of Functional Products and Bioavailability of Phenolic Compounds

**Cristina Monica Dabulici [1], Ionela Sârbu [2] and Emanuel Vamanu [1,***]

[1]   Faculty of Biotechnology, University of Agronomic Science and Veterinary Medicine, 59 Marasti blvd, 1 district, 011464 Bucharest, Romania; dcm.cristina.cm@gmail.com
[2]   Department of Genetics, ICUB-Research Institute of the University of Bucharest, 36-46 Bd. M. Kogalniceanu, 5th District, 050107 Bucharest, Romania; ionela24avram@yahoo.com
*   Correspondence: email@emanuelvamanu.ro; Tel.: +40-742-218-240

Received: 2 July 2020; Accepted: 16 July 2020; Published: 18 July 2020

**Abstract:** The expression of bioactivity depends on the assimilation of different classes of natural substances (e.g., phenolic compounds) in vivo. Six functional extracts (*Aspalathus linearis*, leaves; *Paullinia cupana*, seeds; *Aristotelia chilensis*, berries; *Ilex paraguariensis*, leaves; *Syzygium aromaticum*, cloves, and wild berries) were analyzed in vitro and in vivo as an alternative to alleviating pathologies associated with oxidative stress (proliferation of cancer cells). The purpose of this research was to evaluate the in vitro and in vivo antioxidant and cytotoxic potential of hydroalcoholic solutions, in addition to the assimilation capacity of bioactive components in *Saccharomyces boulardii* cells. In vivo antioxidant capacity (critical point value) was correlated with the assimilation ratio of functional compounds. The results of in vitro antioxidant activities were correlated with the presence of quercetin (4.67 ± 0.27 mg/100 mL) and chlorogenic acid (14.38 ± 0.29 mg/100 mL) in *I. paraguariensis*. Bioassimilation of the main nutraceutical components depended on the individual sample. Phenolic acid levels revealed the poor assimilation of the main components, which could be associated with cell viability to oxidative stress.

**Keywords:** yeast; antioxidant; cytotoxicity; bioavailability; viability

---

## 1. Introduction

Most phenolic compounds are assimilated at the time of transit to the small intestine [1], but some are biotransformed in the colon under the action of the microbiota. As a result, the biological value may be altered because of new compounds resulting after the fermentative action of the microbial pattern. Some portion of these compounds may be completely degraded, as observed for compounds with a high molecular weight (e.g., curcumin) [2]. Despite being known for their antimicrobial effect, certain compounds (such as phenolics) can be assimilated (bioaccumulation) by certain yeast strains, such as *Rhodotorula mucilaginosa* [3]. This bioavailability process has been partially observed for gallic acid. Its use as a carbon source explains the biological response determined by the consumption of these functional molecules [4]. The determination of bioavailability [5] is much more significant in vivo as an indicator of cell absorption because for phenolic compounds the accumulation is difficult to evaluate [6].

The use of a carrier vehicle for the delivery of phenolic compounds increases the in vivo bioavailability [7] and reduces the yield of biotransformations that limit the bioactivity expression. For example, gallic acid has no stability in the fermentative action of the microbiota, but resistance to the action of oxidative stress is mediated through this compound [8]. Its action is limited by the amount that is absorbed into the small intestine, because the remaining amount is inaccessible in vivo [9].

The use of probiotic yeast biomass (such as *Saccharomyces boulardii*) for assimilating these compounds increases bioavailability, and the biological action may be performed in such cases via direct action on the microbial pattern in the colon [2,10]. This process is equivalent to biotransformation into an intermediate compound. Modulation of the microbiological response can be achieved both by restoring eubiosis and promoting the synthesis of compounds with a biomarker role (e.g., short-chain fatty acids) [4]. The metabolism of such compounds has demonstrated the presence of strains belonging to the genus *Bacteroides* and/or the phylum Firmicutes, while the synthesis of compounds has had positive effects on human well-being [11].

New sources of biologically-active compounds have been identified after the determination of bioactivity and the chemical characterization of the natural substrate represented by medicinal herbs [12]. The quantity of functional compounds (such as polyphenolcarboxylic acids) depends on the vegetal material [13], the product conditioning mode, the climatic conditions, the soil composition, and the vegetative stage of the plant [14]. The presence of different bioactive compounds in a large quantity has been correlated with increased resistance to the action of free radicals, which increases life expectancy [15].

A comparison of the biological capacities (such as antioxidant and cytotoxic activities) of different products provides a valid tool to determine their health benefits [16]; functional products have greater benefits because of their antioxidant effect [17]. However, the total antioxidant capacity is the result of the presence of several classes of compounds (such as vitamin C, flavonoids, phenolic acids, and anthocyanins), which affect human health differently. Therefore, in vitro determination does not reflect all of the biological activity [18]. The contribution of individual compounds and/or effects in vivo should be examined through models that provide an image of the interaction and bioassimilation of these molecules [19]. Bioassimilation could be defined as the results of uptake of the bioaccessible phenolic fraction by the yeast cells by mechanisms of transmembrane absorption [20].

Therefore, this study aimed to compare the antioxidant and cytotoxic potentials of hydroalcoholic extracts of the following: (1) *Aspalathus linearis* (rooibos; high in antioxidants, but low tannin content and without caffeine; used for protection against stroke, heart disease, and cancer) [21,22]; (2) *Paullinia cupana* (guarana: used for physical, mental, and cancer-related fatigue with no significant side effects) [23]; (3) *Aristotelia chilensis (Mol.)* Stuntz (maqui; different therapeutic properties based on levels of antioxidants) [24]; (4) *Ilex paraguariensis* A. St.-Hil. (yerba mate; source of biological compounds for the nutraceutical industry) [25]; (5) *Syzygium aromaticum* (L.) Merr. & L.M.Perry (cloves; food preservative and different medicinal purposes) [26]; and (6) wild berry (*Rosa canina*, *Rubus idaeus*, and *Vaccinium myrtillus*—1:1:1; which are relevant to the prevention of degenerative diseases) [27]. The effect on probiotic yeast, *S. boulardii*, was examined by determining the assimilation capacity of cells (enriched biomass) as their in vivo bioassimilation and antioxidant capacity.

## 2. Materials and Methods

### 2.1. Samples

Six dried samples (*A. linearis*, leaves; *P. cupana*, seeds; *A. chilensis*, berries; *I. paraguariensis*, leaves; *S. aromaticum*, cloves, and wild berries) were tested and selected because of their use as alternatives for alleviating pathologies associated with oxidative stress. All samples were purchased from local shops (e.g., Kaufland, Romania) and were dried in an air oven (Memmert oven UFB400) at 40 °C until a constant weight was obtained. The samples (10 g) were mixed with 50% ethanol (to determine a balance between the bioactive compounds present in the extracts) and stored for 48 h in the dark until the mixture was filtered under vacuum. A quantity of 100 g for each sample and 1000 mL solvent was used. The solutions were stored in brown bottles [8].

### 2.2. Determination of Bioactive Compounds

The experiments were performed with high-pressure liquid chromatograph (HPLC), ELITE-LaChrom (Merck-Hitachi, Tokyo, Japan), with DAD (Diode-Array Detection) detector analytic scales (KERN 770). The chromatographic column of stainless steel comprised the stationary stage of octadecylsilane (Inertsil ODS-3 250 * 4.6 mm * 5 μm); mobile stages—mobile stage A, phosphoric acid/water, pH = 2.5 and mobile stage B, methanol. Flow rate of the mobile stage was 1.0 mL/min; elution type: with linear composition gradient of the mobile stage. UV detection was $\Lambda$ = 330 nm; temperature of column oven, 40 °C; injection volume, 20 μL. After the chromatographic system was balanced, the basic line was a straight line and the reference solution was injected. The differences between successive determinations were limited to a maximum of 2%. After the injection of the test solutions, we registered the chromatograms.

Reference solutions (mixtures of 10 μg/mL), were chlorogenic acid, caffeic acid, coumaric acid, ferulic acid, rosemarinic acid, vanillic acid, luteolin 7-glycoside, rutin, apigenin 7-glycoside, luteolin, apigenin, quercetin, quercetin 3-rhamnoside, kaempferol, kaempferol 3-rhamnoside, kaempferol 3-rutinoside, kaempferol 7-glucoside, kaempferol 3-galactoside, catechin, myricetin, myricetin 3-glucoside, and pyrogallol [28,29]. Other reagents used in the determinations were methanol, orthophosphoric acid, ultrapure water, absolute ethanol, and solvent ethanol/water (50:50, *v/v*). For the sample solution, in a measuring bottle of 50 mL, we introduced 0.5000 g of sample powder, adding around 40 mL of solvent, and performing ultrasonication for 30 min at 40 °C. Supplements to 50 mL with the same solvent and filters were added.

The content in the compounds of interest was calculated using the formula compound % = [($A_p$ × $C_e$) × $A_e$] × (50/G) × 100, where $A_p$ is the range of the compound's peak "i" in sample solution, $A_e$ is the range of the compound's peak "i" in the reference solution, $C_e$ is the concentration of compound "i" in the reference solution (mg/mL), G is the quantity of processed sample (mg), and 50 is the correction coefficient [28,29].

### 2.3. Assessment of the In Vitro Antioxidant Potential

The antioxidant potential of all hydroalcoholic extracts was assessed by determining the DPPH (2,2-diphenyl-1-picrylhydrazyl) scavenging activity [30] and the chelating activity [31].

In the case of DPPH scavenging activity, the reaction mixture was 0.1 mL sample, 0.49 mL ethanol, and 0.39 mL DPPH (1 mM). The mixture was kept for 30 min in a dark place. The absorbance of the final mixture was measured at 517 nm against the blank.

In the case of chelating activity, the reaction mixture was 1 mL of sample mixed with 50 μL of 2 mM FeCl₂. After 5 min, 0.2 mL of 5 mM ferrozine solution was added. The mixture was kept at room temperature for 10 min. The absorbance of the final mixture was measured at 562 nm against the blank.

Ascorbic acid (1 mg/mL) and EDTA (Ethylenediaminetetraacetic acid, 0.5 mg/mL) were used as controls, and the remaining activities were calculated using the following formula: $\frac{Abs_M - Abs_P}{Abs_M}$ × 100 [32], where $Abs_M$ is the absorbance of the control and $Abs_P$ is the absorbance of the sample.

### 2.4. Determination of the In Vivo Antioxidant Potential

The antioxidant potential in vivo was evaluated by a modified protocol [2]. A strain of *Saccharomyces boulardii*, a probiotic yeast obtained from the University of Lille, Lille, France, was used. The biomass was obtained using YPG medium (Yeast Extract–Peptone–Glucose; 2% glucose, 2% peptone, and 1% yeast extract) and further cultivated in a lab shaker incubator at 30 °C, for 48 h, at 150 rpm. The yeast cells were separated through centrifugation (4500× *g*, 5 min). The reaction mixture was 0.1 mL sample, 0.1 mL yeast cells in a sterile saline solution, and 0.2 mL H₂O₂. The mixture was made in 100-well honeycomb microplates (sterilized by gamma radiation) and incubated at 30 °C for 1 h using Bioscreen C MBR (Oy Growth Curves Ab Ltd., Helsinki, Finland). The critical concentration was defined as the cross between the viability and mortality lines (different concentrations of H₂O₂ (%)—0.05, 0.1, 0.3, 0.5, 1, 2, and 3), and expressed as a critical point (%). The untreated sample was used as a control for critical point determination. The tests were made until and after bioactive compound assimilation. Finally, the antioxidant potential in vivo was quantified as a percentage value that resulted after the bioactive compound assimilation. The results were calculated in comparison with a control that was realized at the same critical point for each sample but without an extract in the reaction mixture.

### 2.5. Bioavailability Index Quantification

The bioavailability index was determined based on the method presented by Celep et al. (2018) [33], with some modifications, using the same *S. boulardii* strain. The extract (1 mg/mL), was added to the culture media via filtration through sterile Milllipore membrane (0.22 μm) and inserted in the temperature-controlled orbital shaker (LabTech) in an 80 mL Duran bottle (sterile-venting membrane screw caps; 0.2 μm (ePTFE membrane). Chlorogenic acid (1 mg/mL) was used as a control. The bioavailability index (BI) was calculated using the following formula: $\frac{\text{quantity of absorbed phenolics}}{\text{quantity of total phenolics}} \times$ 100 [2,33], where the quantity of absorbed phenolics was estimated after the biomass was freeze-dried and the presence of total phenolics was determined as presented in Section 2.2.

### 2.6. Evaluation of Cytotoxicity

The cytotoxic effect of hydroalcoholic extracts was assessed by measuring HCT-8 and *S. boulardii* cell viability using the Vybrant® MTT (3-(4,5-dimethylthiazol-2-yl)-2,5-diphenyltetrazolium bromide) Cell Proliferation Assay Kit. The human epithelial HCT-8 cell line isolated from ileocecal colorectal adenocarcinoma, passage 18, was cultivated in RPMI 1640 (Sigma, St. Louis, MO, USA) supplied with 10% fetal bovine serum (FBS; Biochrom, Berlin, Germany) in polystyrene 96-well plates at 37 °C, 5% $CO_2$ until they reached a 60% confluence; *S. boulardii* strain was cultured in YPG at 30 °C for 24 h, 120 rpm.

The medium was removed and the cells were incubated for 24 h in fresh media with extracts in concentrations of 10% and 1%. After the incubation, the medium was removed and the cells were washed one time with warm PBS (phosphate-buffered saline) and incubated with MTT solution for 2.5 h (human cell line) and 4 h (yeast strain).

The dye was solubilized with DMSO (Dimethyl sulfoxide) and the plate was read at 540 nm using Synergy HTX (Biotek, Winooski, VT, USA). All of the samples were performed in triplicate. Cell viability was calculated according to the formula: % cell survival = (mean sample absorbance/mean control absorbance) × 100. An untreated sample with extracts (only with 5% and 0.5% ethanol, respectively) was used as control [34].

### 2.7. Statistical Analysis

All of the parameters investigated were evaluated in a minimum of three independent determinations, and the results were expressed as the mean ± standard deviation (SD). The mean and SD values were calculated using the IBM SPSS Statistics 23 software package (IBM Corporation, Armonk, NY, USA). The significance level for the calculations was set as follows: significant, $p \leq 0.05$; very significant, $p \leq 0.01$; and highly significant, $p \leq 0.001$, using the normal distribution of the variables. The differences were analyzed by ANOVA followed by a Tukey post hoc analysis. Analysis and correlation of the experimental data were conducted with the IBM SPSS Statistics software package (IBM Corporation, Armonk, NY, USA) [35].

## 3. Results

### 3.1. Determination of Bioactive Compounds

The ability of some nutraceuticals to reduce the physiological effects of oxidative stress was related to the distribution of bioactive compounds (like phenolic acids and flavonoids). Thus, Figure 1 depicts the level of major compounds responsible for expressing the biological effects. The complex distribution of nutraceutical effect molecules was evidenced in the *S. aromaticum* extract (5732.65 ± 87.59 μg/mL gallic acid equivalent), with significant differences from the remainder of the extracts ($p \leq 0.001$). For *A. linearis* and maqui, the total phenolic quantity was similar to the wild berries sample. In contrast, the reduced quantity of phenolic compounds present when using *P. cupana* was inversely proportional to the expression of the antioxidant capacity in vitro [36]. In the case of the total flavonoid content measurement (Figure 2), the result was different; this result could be interpreted as an effect determined primarily by the tested samples and solvent. This was the case of *P. cupana* extract, which was similar to that of *S. aromaticum* (approximately 7.00 µg/mL, $p \leq 0.01$) [37].

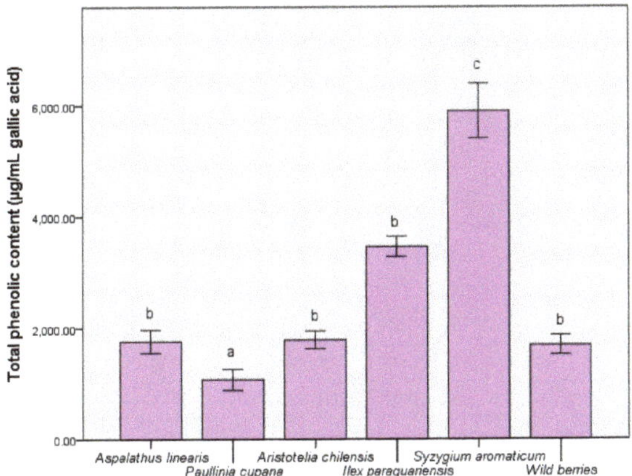

**Figure 1.** The total quantities of phenolic compounds in the extracts. Different letters in each column represent significant statistical differences ($p \leq 0.05$) between extracts, n = 4.

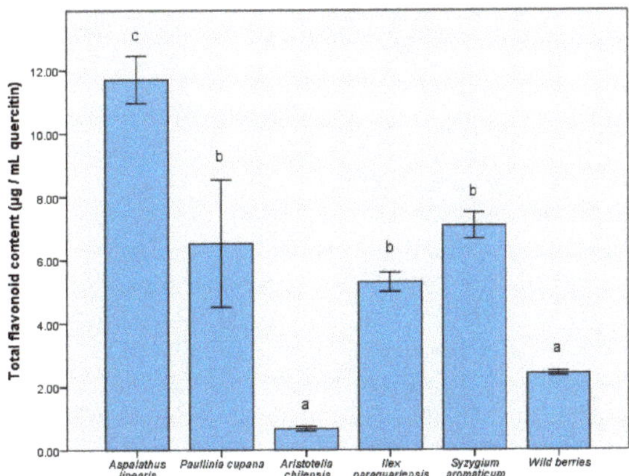

**Figure 2.** The total quantities of flavonoid compounds in the extracts. Different letters in each column represent significant statistical differences ($p \leq 0.05$) between extracts, n = 4.

A similar pattern of major bioactive compounds was identified in the test samples, in which the main phenolic acids were chlorogenic, caffeic acids, and quercetin, as a flavonoidic compound. Table 1 reveals a high number of flavonoids and a varied distribution of kaempferol derivatives. The in vitro response was an expression of the distribution of these compounds [38], particularly for *S. aromaticum* and *I. paraguariensis*. In addition, the quercetin level for *S. aromaticum* was 4.67 ± 0.27 ($p \leq 0.001$), greater than phenolic acids. Chlorogenic acid contained in the extract of *I. paraguariensis* showed a higher level compared to that in the remaining samples and control ($p \leq 0.001$). The difference between the chemical quantification (Table 1) and assay results (Figure 1) was determined by the possible presence of other compounds in trace amounts that could not be determined. The solvent used (50% ethanol) determined a high solubility in water and the differences between these two assays were a result of this aspect. They were in quantities that did not directly affect the bioavailability results [2].

**Table 1.** Quantitative analysis of major bioactive compounds from the extracts.

| Compounds (mg/100 mL Extract) | *Aspalathus linearis* | *Paullinia cupana* | *A. chilensis* | *Ilex paraguariensis* | *Syzygium aromaticum* | **Wild Berries** |
|---|---|---|---|---|---|---|
| Chlorogenic acid | 0.21 ± 0.01 [a] | 0.60 ± 0.01 [a] | 0.04 ± 0.01 [a] | 14.38 ± 0.29 [c] | 0.06 ± 0.01 [a] | 0.37 ± 0.01 [a] |
| Caffeic acid | 0.11 ± 0.01 [a] | 0.01 ± 0.01 [a] | 0.005 ± 0.02 [a] | 0.19 ± 0.10 [a] | 0. 60 ± 0.08 [b] | 0.30 ± 0.01 [b] |
| Rutin | nd | nd | nd | 0.13 ± 0.01 [a] | 0.11 ± 0.01 [b] | 0.07 ± 0.08 [b] |
| Quercetin | 0.68 ± 0.04 [c] | 0.20 ± 0.01 [a] | 0.13 ± 0.01 [a] | 0.96 ± 0.04 [c] | 4.67 ± 0.27 [c] | 0.21 ± 0.02 [b] |
| Kaempferol 7-glucoside | 0.04 ± 0.01 [c] | 0.06 ± 0.05 [b] | 0.04 ± 0.01 [b] | 2.88 ± 0.30 [c] | 0.02 ± 0.01 [a] | 0.02 ± 0.01 [a] |
| Kaempferol 3-galactoside | 0.02 ± 0.03 [b] | 0.24 ± 0.01 [a] | 0.33 ± 0.05 [c] | 6.09 ± 0.29 | 1.07 ± 0.05 [c] | 0.01 ± 0.03 [c] |
| Kaempferol 3-rutinoside | 0.06 ± 0.02 [a] | 0.03 ± 0.05 [a] | 0.04 ± 0.02 [a] | 0.56 ± 0.41 | 0.10 ± 0.07 [b] | 0.03 ± 0.01 [a] |
| Kaempferol 3-rhamnoside | 0.01 ± 0.01 [a] | 0.02 ± 0.05 [a] | nd | 0.24 ± 0.03 [b] | 0.04 ± 0.08 [c] | 0.08 ± 0.05 [b] |
| Myricetin 3-glucoside | 0.95 ± 0.08 [b] | 0.11 ± 0.01 [a] | 0.09 ± 0.04 [b] | 0.03 ± 0.01 [b] | 0.40 ± 0.03 [b] | 0.002 ± 0.07 [a] |
| Quercetin 3-rhamnoside | 0.14 ± 0.01 [a] | 1.84 ± 0.04 [c] | 0.31 ± 0.09 [b] | 40.52 ± 0.93 | 0.37 ± 0.01 [b] | 0.24 ± 0.01 [a] |

Different letters represent significant statistical differences ($p \leq 0.05$) between extracts, n = 3; nd—not detected.

*3.2. Determination of In Vitro Antioxidant Activities*

We also sought to determine the antioxidant potential expressed as the DPPH-scavenging activity and chelation capacity (Figure 3). The maqui extract demonstrated the highest scavenging activity (81.71 ± 2.73%), which was similar to that of ascorbic acid (as a control), and 25% higher than that of the control sample ($p \leq 0.001$). The remainder of the extracts showed values within ±10% of the second blank. The overall data suggested that maqui is a potential source of antioxidant compounds, the results being sustained by a previous study that demonstrated that the freeze-dried samples retained the highest quantity of bioactive compounds [39].

*I. paraguariensis* demonstrated a chelation capacity at least 25% higher than that of the control (88.47 ± 1.25%). The results demonstrated a prophylactic role in preventing the generation of free radicals in the case of administration of extracts as functional supplements. In addition, *A. linearis* exhibited a balance of the two properties (65.90 ± 1.65% for DPPH scavenging activity and 66.95 ± 1.39% for chelating activity), demonstrating pharmacological importance through multiple actions.

The exploitation of bioactive components in extracts also consists of their assimilation by biomass and the increase in the resistance to oxidative stress. This study aimed to use *S. boulardii* as a model for demonstrating the stability and resistance of eukaryotic yeast to the presence of free radicals.

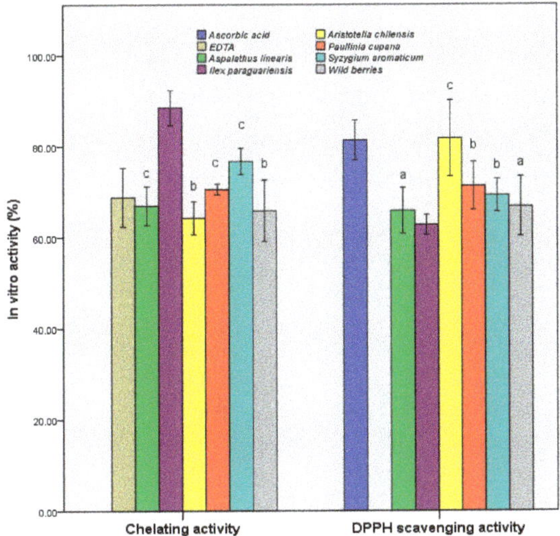

**Figure 3.** In vitro antioxidant activities of the samples. Different letters represent significant statistical differences (EDTA/ascorbic acid vs. samples; $p \leq 0.05$), n = 3; EDTA/ascorbic acid was used as control.

### 3.3. Determination of Bioavailability Index

According to Figure 4, the highest BI was determined for *I. paraguariensis* extract, with an average value exceeding 60% ($p \leq 0.05$). Wild berries showed the smallest value ($p \leq 0.01$ vs. chlorogenic acid), although they are currently a widely-used product. The BI value was similar to the presence of maqui. It was possible to identify the correlation between the total presence of bioactive components, particularly chlorogenic acid. The stability of this compound in fermented medium is an advantage over the remainder of the unstable components (e.g., quercetin). This result showed that it cannot be used as a carbon source in the presence of probiotic strains, such as *S. boulardii*. These results support findings of recent studies that demonstrated the significant role of chlorogenic acid in the control of oxidative-stress-related causes [40].

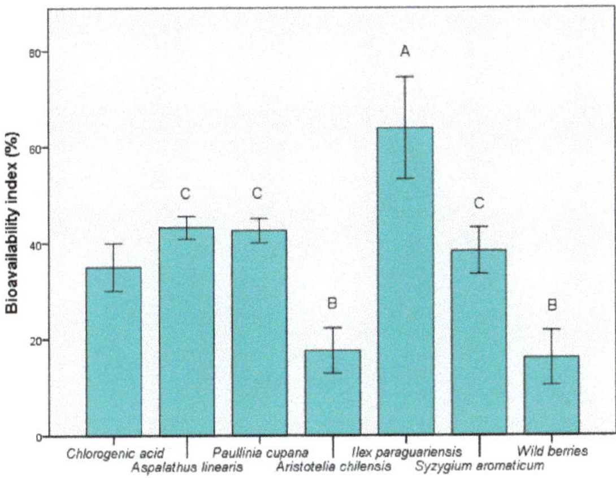

**Figure 4.** Bioavailability index of extracts based on *S. boulardii* in vivo model. Different letters represent significant statistical differences (chlorogenic acid vs. samples; $p \leq 0.05$), n = 3; chlorogenic acid was used as control.

This compound is important in its action in the colon because bioavailability depends on the metabolism exerted by the microbiota. It can be absorbed directly without affecting its chemical structure in case it directly reaches the colon. Otherwise, its derivatives (caffeic acid) are responsible for the antioxidant effects in vivo [41], and they can protect the eukaryotic cells in the presence of free radicals.

### 3.4. Determination of In Vivo Antioxidant Activities after the Assimilation Process

In Figure 5, the in vivo effect of the presence of the functional components is shown by the critical point value. The content of phenolic compounds showed a direct correlation, although for maqui, *S. aromaticum*, and *I. paraguariensis* the values were not significantly different ($p < 0.01$). A direct relationship between bioavailability index (Figure 4) and critical point value (Figure 5) was determined for *A. linearis* and *P. cupana*. For these tests, the results demonstrated a balance between exo- and endo-protection offered by these extracts. The exception was the use of *I. paraguariensis* extract which resulted in a highly-critical point, $1.40 \pm 0.01\%$ ($p < 0.01$ vs. untreated sample). This result demonstrated that a high quantity of phenolic compounds does not always cause in vivo protection against oxidative stress.

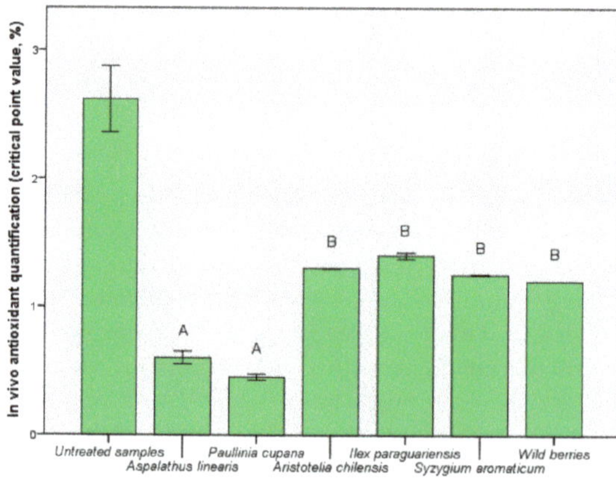


**Figure 5.** In vivo antioxidant activity before the bioassimilation process, but in the presence of extracts. Different letters represent significant statistical differences (untreated sample vs. samples; $p \leq 0.05$), n = 3; untreated sample was used as control.

After the bioactive compound assimilation, the improvement of viability is shown in Figure 6. The extract of *P. cupana* caused the loss of viability, and the result was considered to be determined by a pronounced antimicrobial effect or its use as a carbon source. Exo-protection calculated for *I. paraguariensis* could not be obtained even after fermentative assimilation. The result was similar to the control sample (chlorogenic acid) but ~60% higher compared to wild berries. The bioactive component was best assimilated for the extract of *S. aromaticum*, offering a viability increase of $21.21 \pm 0.03\%$ compared to the exo-protection phase. The result was interpreted as an effect of quercetin assimilation (Table 1).

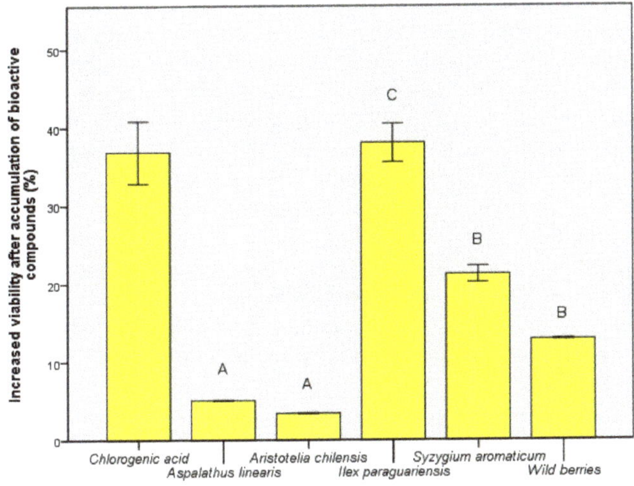

**Figure 6.** Increased viability of *S. boulardii* cells after bioactive compound assimilation. Different letters represent significant statistical differences (chlorogenic acid vs. samples; $p \leq 0.05$), n = 3; chlorogenic acid was used as control.

## 3.5. Determination of Human Cell Viability

The human epithelial cell viability in the presence of the plant extracts is shown in Figure 7. Exposure of the epithelial cells at low concentrations of hydroalcoolique plant extracts (1%) did not affect their viability; the exception was *S. aromaticum*, with an increase of cell viability of 10%. Instead, exposure of the cell line HCT-8 to higher concentrations of wild fruits and *I. paraguariensis* had a strong cytotoxic effect. The MTT assays measured the mitochondrial activity of cells. There was the possibility that *S. aromaticum* stimulated the cell enzyme activity. Other factors might have determined higher cell proliferation, such as slightly more cells were inoculated due to small pipetting errors, a favorable position in the plate, or just natural variations of cellular metabolism.

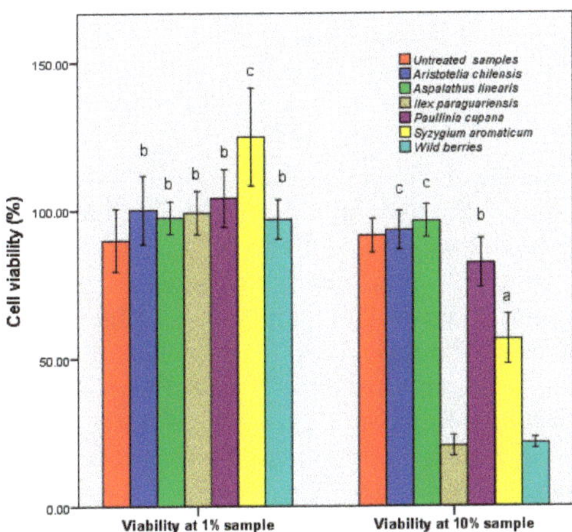

**Figure 7.** Evaluation of the HCT-8 cell viability via MTT assay in the presence of hydroalcoholic extracts. Different letters represent significant statistical differences (untreated sample vs. samples; $p \leq 0.05$), n = 3; untreated sample was used as control.

The cytotoxic activity was high in the extracts where the rutin was absent (Table 1), and the bioavailability showed average values. The minimum obtained was recorded in the samples that contained high amounts of phenolic acids, which could be understood as a form of exo-protection. Low flavonoid concentration determined specificity, which corresponded to previous studies performed on other tumor and nontumor cells [42]. For *A. linearis* the balanced distribution of the two major phenolic acids corresponded to a maximum cell number. The obtained results showed that the chemical composition was the main factor that influenced the cytotoxic activity. The HCT-8 cells morphology modifications after cultivation with 10% extracts are presented in Supplementary Figure S1.

The metabolic activity of the strain *S. boulardii* was modulated uniquely by plant extracts. Wild berries, in addition to *A. linearis* had an inhibitory effect (Figure 8). In contrast, *P cupana* stimulated yeast cell metabolism ($p \leq 0.001$). The effect of wild berries confirmed the cytotoxic effect shown at 10% concentration. This aspect demonstrated that treatments with wild berries and *I. paraguariensis* affect the metabolic activity of both yeast and human epithelial cell line HCT-8.

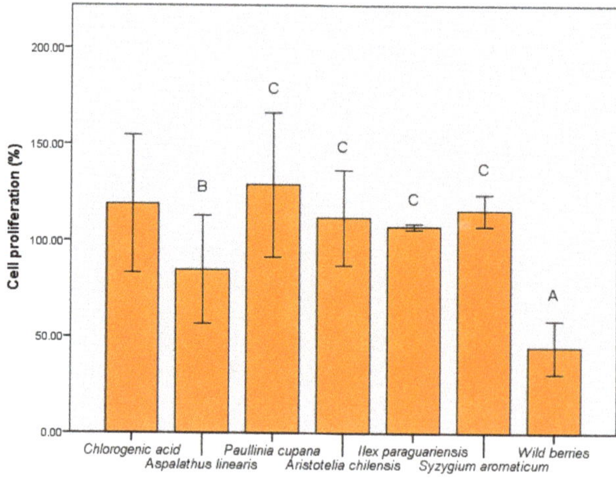

**Figure 8.** MTT assay of *S. boulardii* cells cultivated in the presence of hydroalcoholic extracts. Different letters represent significant statistical differences (chlorogenic acid vs. samples; $p \leq 0.05$), n = 3; chlorogenic acid was used as control.

## 4. Discussion

The relationship between the degree of protection expressed in vitro and the possibility of any in vivo action was affected by the high bioassay variability as an expression of the biomass biological value. From the data obtained we can consider that such biomass (for example, for *I. paraguariensis*) provides protection from the formation of free radicals that cause oxidative stress. This study showed that bioaccessibility and bioavailability are indicators of the in vitro/in vivo relationship. The identification of the critical point of contact led to the definition of innovative mechanisms for the exploitation of bioactive components that can ultimately improve the bioaccessibility and bioavailability of the target compound. This aspect should be confirmed by further experiments. This in vivo indicator expressed the potential of the product (such as a functional extract, *I. paraguariensis*) to sustain the physiological mechanisms of protection against oxidative stress. This was an image of biological action after administration and could determine a real characterization of the bioactivity and bioavailability (Figures 4 and 5).

The ability to assimilate yeast cells led to an understanding of the stability of the functional components and the importance that the pattern and the level of these compounds play in vivo. Thus, one of the perspectives that this study demonstrated was that increasing the concentration of a compound is not the essential factor; rather, it is the distribution of the whole spectrum of nutraceuticals in the tested product. This aspect explains the negative effects of the administration of green tea extracts [43] or the lack of a clinical result in many dietary supplements based on medicinally- or nutritionally-functional compounds [44]. The correlation between in vitro and in vivo data must be validated using laboratory models (yeasts) to demonstrate that a nutraceutical can also represent a pharmaceutical product. This link is one of the limiting points in the development of competitive products for personalized treatments (reducing inflammatory processes, for example) [2]. Such results would clinically validate the expansion of the nutraceutical industry to limit the action of the factors that determine degenerative pathologies.

The results shown in Figure 4 illustrate the high degree of absorption of the bioactive components of *I. paraguariensis* and *A. linearis*. The use of yeast was seen as an in vivo model to demonstrate the ability of the phenolic component for absorption after oral administration. This property was not correlated with cytotoxic capacity, demonstrating the biological versatility of chlorogenic acid as a bioactive compound. This study has contributed to the understanding of how the phytocomplex (particularly the active compounds) influences the eukaryotic cell response to oxidative stress. A chemical versatility that was controlled by the type of cells it interacted with was also demonstrated. The current study has shown that the phenolic pattern influences the same type of cells differently. The increase in the bioavailability index was correlated with poor cytotoxicity for *I. paraguariensis*.

Although the study did not take into account the influence of the physico-chemical factors during the gastrointestinal transit, it was considered that the remainder of the samples do not have a major influence as protectors in the action of oxidative stress. It can be assumed that the absorption in *S. boulardii* follows the same general principles as the absorption in the human intestinal lumen (e.g., molecular mass and degree of biotransformation of the phenolic component). Without eliminating comparative differences with multicellular organisms, strains belonging to the genus *Saccharomyces* are considered to be a model organism that can provide an overview of pharmacological applications that could be developed against oxidative stress [45].

Thus, the pattern of functional compounds and their molar ratio was considered the primary cause of bioavailability following prolonged exposure. The cytotoxic activity was high (i.e., over 90%) in extracts where rutin was absent (Table 1) because of possible bioconversion into rutin sulphate (metabolite; [46]). In addition, xenobiotic effects are possible as a secondary cause of blocking the cellular uptake process. Such effects were determined by secondary compounds, which act on the membrane transport mechanisms. In some studies, the effect of cell non-accumulation was considered as a positive effect (in the liver) because it did not involve increased cytotoxicity.

This study presented a series of in vitro and in vivo data that explain the exo- and endo-cellular mode of action of functional extracts. The biological effect showed specificity based on the profile of the bioactive compounds. It has been shown that different phenolic compounds have a characteristic ability to be assimilated, which explains the value of bioassimilation. The high capacity of incorporation of quercetin into eukaryotic cells increased the degree of protection against oxidative stress (Figures 5 and 6). Phenolic acids (such as chlorogenic acid) did not significantly influence endo-oxidative protection. A positive correlation between phenolic compounds and in vitro activities was calculated for all extracts, but *I. paraguariensis* and *S. aromaticum* presented low values ($r \approx 0.3$), similar to that of wild berries. A similar value was obtained in the case of cytotoxic activity of the same extracts and results observed for other berry samples, not only for representatives of the same berry. A high correlation value was calculated for *A. linearis* ($r \geq 0.5$), and also for *P. cupana* and maqui ($r \geq 0.9$) in the case of both in vitro activities. In contrast, in vivo antioxidant protection was directly correlated with the presence of the phenolic component ($r \geq 0.9$) for all extracts.

This fact confirms that this acid is relevant as a non-pharmacological product because it has a non-invasive character in the prevention of degenerative pathologies [47]. For quercetin, the biological response was characteristic of a certain microbial species. Species of the same genus may have variability in the case of the bioavailability–biological effect. The results were confirmed by clinical studies that concluded that this compound showed differences at the individual level characterized by a pharmacologically-variable bioefficiency [48].

The bioavailability and expression of antioxidant action are dependent upon their primary action, and, by oral consumption, the function is profoundly affected in vivo [49]. Bioavailability via enriched biomass is a useful method for increasing the pharmacological value of *S. boulardii*, which enhances the assimilation of the active substance in the human colon. This is important because it increases the resistance of probiotic biomass to oxidative stress present in the case of dysbiosis. These products will maintain greater viability when interacting with the target microbiota and will be able to exert their effect for a longer time, an essential property in the microbial modulation process. In vivo delivery of compounds may be considered to improve blood concentration without considering biotransformations that may occur along the gastrointestinal tract. These aspects were supported by the present results, providing a lab mirror of the functional product that will have an in vivo effect.

## 5. Conclusions

In addition to the increase in the oxidative stress stability, bioassimilation and bioavailability of phenolic compounds define the in vivo complex action that phenolic compounds exert on interaction with eukaryotic cells. In conclusion, the use of *I. paraguariensis* extract increased the cellular capacity to protect against exogenous factors supporting oxidative stress compared with both the control sample and the assimilation of *I. paraguariensis*. The level and pattern of phenolic compounds expressed the antioxidant and cytotoxic activities. A correlation between bioactive compounds and different activities in vitro/in vivo was not obtained in the case of high levels because it was a species-specific relationship. An accurate understanding of bioassimilation and bioavailability processes will define the ability of probiotic biomasses to release functional components and will improve nutraceutical formulation.

**Supplementary Materials:** The following are available online at http://www.mdpi.com/2304-8158/9/7/953/s1, Figure S1: HCT-8 cells' morphology modifications after cultivation with 10% extracts, 200X.

**Author Contributions:** C.M.D., E.V., and I.S. designed the experiments, analyzed the data, and wrote the paper. The authors discussed and made comments on the results. All authors have read and agreed to the published version of the manuscript.

**Funding:** The support for the study was obtained by equal contributions of the authors. It was partially supported by UASVM Bucharest, Romania's doctoral studies.

**Acknowledgments:** Many thanks to Sultana Niță, who helped us perform chromatographic analysis (National Institute of Chemical–Pharmaceutical Research and Development—Department of Physical-Chemical Analysis and Quality Control), and Constantin Erena (ErenaMed Clinic SRL Bucharest, Romania) for samples.

**Conflicts of Interest:** The authors declare no conflict of interest.

## References

1. Tarko, T.; Duda-Chodak, A.; Sroka, P.; Satora, P.; Michalik, J. Transformation of Phenolics in Alimentary Tract. *Food Technol. Biotechnol.* **2009**, *47*, 456–463. Available online: https://pdfs.semanticscholar.org/58f2/23fafbe920b79ce4dbff4ec194e94b33007e.pdf (accessed on 1 June 2020).

2. Vamanu, E.; Gatea, F.; Sârbu, I.; Pelinescu, D. An In Vitro Study of the Influence of Curcuma longa Extracts on the Microbiota Modulation Process, In Patients with Hypertension. *Pharmaceutics* **2019**, *11*, 191. [CrossRef]

3. Jarboui, R.; Baati, H.; Fetoui, F.; Gargouri, A.; Gharsallah, N.; Ammar, E. Yeast performance in wastewater treatment: Case study ofRhodotorula mucilaginosa. *Environ. Technol.* **2012**, *33*, 951–960. [CrossRef] [PubMed]

4. Vamanu, E. Polyphenolic Nutraceuticals to Combat Oxidative Stress Through Microbiota Modulation. *Front. Pharmacol.* **2019**, *10*, 492. [CrossRef] [PubMed]

5.  Aqil, F.; Munagala, R.; Jeyabalan, J.; Vadhanam, M.V. Bioavailability of phytochemicals and its enhancement by drug delivery systems. *Cancer Lett.* **2013**, *334*, 133–141. [CrossRef] [PubMed]

6.  D'Archivio, M.; Filesi, C.; Varì, R.; Scazzocchio, B.; Masella, R. Bioavailability of the Polyphenols: Status and Controversies. *Int. J. Mol. Sci.* **2010**, *11*, 1321–1342. [CrossRef] [PubMed]

7.  Vulić, J.; Šeregelj, V.; Kalušević, A.; Lević, S.; Nedović, V.; Tumbas Šaponjac, V.; Čanadanović-Brunet, J.; Ćetković, G. Bioavailability and Bioactivity of Encapsulated Phenolics and Carotenoids Isolated from Red Pepper Waste. *Molecules* **2019**, *24*, 2837. [CrossRef]

8.  Vamanu, E.; Gatea, F.; Sârbu, I. In Vitro Ecological Response of the Human Gut Microbiome to Bioactive Extracts from Edible Wild Mushrooms. *Molecules* **2018**, *23*, 2128. [CrossRef]

9.  Hussain, M.B.; Hassan, S.; Waheed, M.; Javed, A.; Farooq, M.A.; Tahir, A. *Bioavailability and Metabolic Pathway of Phenolic Compounds, Plant Physiological Aspects of Phenolic Compounds*; IntechOpen: London, UK, 2019. [CrossRef]

10.  Vamanu, E.; Gatea, F. Correlations between Microbiota Bioactivity and Bioavailability of Functional Compounds: A Mini-Review. *Biomedicines* **2020**, *8*, 39. [CrossRef]

11.  Esteban-Torres, M.; Santamaría, L.; Cabrera-Rubio, R.; Plaza-Vinuesa, L.; Crispie, F.; De las Rivas, B.; Cotter, P.; Muñoz, R. A Diverse Range of Human Gut Bacteria Have the Potential To Metabolize the Dietary Component Gallic Acid. *Appl. Environ. Microbiol.* **2018**, *84*, e01558-18. [CrossRef]

12.  Tyśkiewicz, K.; Konkol, M.; Kowalski, R.; Rój, E.; Warmiński, K.; Krzyżaniak, M.; Gil, Ł.; Stolarski, M.J. Characterization of bioactive compounds in the biomass of black locust, poplar and willow. *Trees* **2019**, *33*, 1235–1263. [CrossRef]

13.  Altemimi, A.; Lakhssassi, N.; Baharlouei, A.; Watson, D.G.; Lightfoot, D.A. Phytochemicals: Extraction, Isolation, and Identification of Bioactive Compounds from Plant Extracts. *Plants* **2017**, *22*, 42. [CrossRef] [PubMed]

14.  Drobek, M.; Frąc, M.; Cybulska, J. Plant Biostimulants: Importance of the Quality and Yield of Horticultural Crops and the Improvement of Plant Tolerance to Abiotic Stress—A Review. *Agronomy* **2019**, *9*, 335. [CrossRef]

15.  Peng, C.; Wang, X.; Chen, J.; Jiao, R.; Wang, L.; Li, Y.M.; Zuo, Y.; Liu, Y.; Lei, L.; Ma, K.Y.; et al. Biology of ageing and role of dietary antioxidants. *BioMed Res. Int.* **2014**, *2014*, 13. [CrossRef] [PubMed]

16.  Patterson, T.L.; Mausbach, B.T. Measurement of functional capacity: A new approach to understanding functional differences and real-world behavioral adaptation in those with mental illness. *Annu. Rev. Clin. Psychol.* **2010**, *6*, 139–154. [CrossRef]

17.  Lourenço, S.C.; Moldão-Martins, M.; Alves, V.D. Antioxidants of Natural Plant Origins: From Sources to Food Industry Applications. *Molecules* **2019**, *24*, 4132. [CrossRef]

18.  Skrovankova, S.; Sumczynski, D.; Mlcek, J.; Jurikova, T.; Sochor, J. Bioactive Compounds and Antioxidant Activity in Different Types of Berries. *Int. J. Mol. Sci.* **2015**, *16*, 24673–24706. [CrossRef]

19.  Papadimitriou, K.; Zoumpopoulou, G.; Foligné, B.; Alexandraki, V.; Kazou, M.; Pot, B.; Tsakalidou, E. Discovering probiotic microorganisms: In Vitro, In Vivo, genetic and omics approaches. *Front. Microbiol.* **2015**, *6*, 58. [CrossRef]

20.  Available online: http://scitechconnect.elsevier.com/bioavailability-bioaccessibility-bioactivity-food-components/ (accessed on 12 June 2020).

21.  Available online: http://abc.herbalgram.org/site/PageServer (accessed on 12 March 2020).

22.  Ajuwon, O.R.; Katengua-Thamahane, E.; Van Rooyen, J.; Oguntibeju, O.O.; Marnewick, J.L. Protective Effects of Rooibos (*Aspalathus linearis*) and/or Red Palm Oil (*Elaeis guineensis*) Supplementation on *tert*-Butyl Hydroperoxide-Induced Oxidative Hepatotoxicity in Wistar Rats. *Evid. Based Complement. Altern. Med.* **2013**, *2013*, 984273. [CrossRef]

23.  Paschoin de Oliveira Campos, M.; Riechelmann, R.; Martins, L.C.; Hassan, B.J.; Branco Assunção Casa, F.; Del Giglio, A. Guarana (*Paullinia cupana*) Improves Fatigue in Breast Cancer Patients Undergoing Systemic Chemotherapy. *J. Altern. Complement. Med.* **2011**, *17*, 505–512. [CrossRef]

24.  Misle, E.; Garrido, E.; Contardo, H.; González, W. Maqui [*Aristotelia chilensis* (Mol.) Stuntz] the Amazing Chilean Tree: A Review. *J. Agric. Sci. Technol. B* **2011**, *1*, 473–482.

25.  Heck, C.I.; De Mejia, E.G. Yerba Mate Tea (*Ilex paraguariensis*): A Comprehensive Review on Chemistry, Health Implications, and Technological Considerations. *J. Food Sci.* **2007**, *72*, 138–151. [CrossRef]

26.  Cortés-Rojas, D.F.; De Souza, C.R.; Oliveira, W.P. Clove (*Syzygium aromaticum*): A precious spice. *Asian Pac. J. Trop. Biomed.* **2014**, *4*, 90–96. [CrossRef]

27. Diaconeasa, Z.; Iuhas, C.I.; Ayvaz, H.; Rugină, D.; Stanilă, A.; Dulf, F.V.; Bunea, A.; Socaci, S.A.; Socaciu, C.; Pintea, A. Phytochemical Characterization of Commercial Processed Blueberry, Blackberry, Blackcurrant, Cranberry, and Raspberry and Their Antioxidant Activity. *Antioxidants* **2019**, *8*, 540. [CrossRef]

28. Fanali, C.; Della Posta, S.; Vilmercati, A.; Dugo, L.; Russo, M.; Petitti, T.; Mondello, L.; De Gara, L. Extraction, Analysis, and Antioxidant Activity Evaluation of Phenolic Compounds in Different Italian Extra-Virgin Olive Oils. *Molecules* **2018**, *23*, 3249. [CrossRef]

29. Vamanu, E.; Nita, S. Antioxidant Capacity and the Correlation with Major Phenolic Compounds, Anthocyanin, and Tocopherol Content in Various Extracts from the Wild Edible *Boletus edulis* Mushroom. *BioMed Res. Int.* **2012**, *2013*, 313905. [CrossRef] [PubMed]

30. Kedare, S.B.; Singh, R.P. Genesis and development of DPPH method of antioxidant assay. *J. Food Sci. Technol.* **2011**, *48*, 412–422. [CrossRef] [PubMed]

31. Raudone, L.; Vilkickyte, G.; Pitkauskaite, L.; Raudonis, R.; Vainoriene, R.; Motiekaityte, V. Antioxidant Activities of *Vaccinium vitis-idaea* L. Leaves within Cultivars and Their Phenolic Compounds. *Molecules* **2019**, *24*, 844. [CrossRef]

32. Shoibe, M.; Chy, M.N.U.; Alam, M.; Adnan, M.; Islam, M.Z.; Nihar, S.W.; Rahman, N.; Suez, E. In Vitro and In Vivo Biological Activities of *Cissus adnata* (Roxb.). *Biomedicines* **2017**, *5*, 63. [CrossRef]

33. Celep, E.; Akyüz, S.; İnan, Y.; Yesilada, E. Assessment of potential bioavailability of major phenolic compounds in *Lavandula stoechas* L. ssp. *Stoechas*. *Ind. Crops Prod.* **2018**, *118*, 111–117. [CrossRef]

34. Mihai, M.M.; Holban, A.M.; Giurcăneanu, C.; Popa, L.G.; Buzea, M.; Filipov, M.; Lazar, V.; Chifiriuc, M.C.; Popa, M.I. Identification and Phenotypic Characterization of the Most Frequent Bacterial Etiologies in Chronic Skin Ulcers. *Rom. J. Morphol. Embryol.* **2014**, *55*, 1401–1408. Available online: http://www.rjme.ro/RJME/resources/files/55041414011408.pdf (accessed on 21 January 2020). [PubMed]

35. Vamanu, E.; Pelinescu, D.; Gatea, F.; Sârbu, I. Altered in Vitro Metabolomic Response of the Human Microbiota to Sweeteners. *Genes* **2019**, *10*, 535. [CrossRef] [PubMed]

36. Bonilla, J.; Sobral, P.J.A. Antioxidant and antimicrobial properties of ethanolic extracts of guarana, boldo, rosemary and cinnamon Propriedades antioxidante e antimicrobiana de extratos etanólicos de guaraná, boldo, alecrim e canela. *Braz. J. Food Technol.* **2017**, *20*, e2016024. [CrossRef]

37. Ghasemzadeh, A.; Jaafar, H.Z.E.; Rahmat, A. Antioxidant Activities, Total Phenolics and Flavonoids Content in Two Varieties of Malaysia Young Ginger (*Zingiber officinale* Roscoe). *Molecules* **2010**, *15*, 4324–4333. [CrossRef] [PubMed]

38. Hughes, J.P.; Rees, S.; Kalindjian, S.B.; Philpott, K.L. Principles of early drug discovery. *Br. J. Pharmacol.* **2011**, *162*, 1239–1249. [CrossRef] [PubMed]

39. Quispe-Fuentes, I.; Vega-Gálvez, A.; Aranda, M. Evaluation of phenolic profiles and antioxidant capacity of maqui (Aristotelia chilensis) berries and their relationships to drying methods. *J. Sci. Food Agric.* **2018**, *98*, 4168–4176. [CrossRef]

40. Wang, Z.; Lam, K.L.; Hu, J.; N Ge, S.; Zhou, A.; Zheng, B.; Zeng, S.; Lin, S. Chlorogenic acid alleviates obesity and modulates gut microbiota in high-fat-fed mice. *Food Sci. Nutr.* **2019**, *7*, 579–588. [CrossRef]

41. Gonthier, M.P.; Verny, M.A.; Besson, C.; Rémésy, C.; Scalbert, A. Chlorogenic Acid Bioavailability Largely Depends on Its Metabolism by the Gut Microflora in Rats. *J. Nutr.* **2003**, *133*, 1853–1859. [CrossRef]

42. Calhelha, R.C.; Falcão, S.I.; Queiroz, M.J.; Vilas-Boas, M.; Ferreira, I.C. Cytotoxicity of Portuguese Propolis: The Proximity of the In Vitro Doses for Tumor and Normal Cell Lines. *BioMed Res. Int.* **2014**, *2014*, 897361. [CrossRef]

43. Hu, J.; Webster, D.; Cao, J.; Shao, A. The safety of green tea and green tea extract consumption in adults—Results of a systematic review. *Regul. Toxicol. Pharmacol.* **2018**, *95*, 412–433. [CrossRef]

44. Royal Pharmaceutical Society. Available online: https://www.pharmaceutical-journal.com/1-what-is-a-nutraceutical/20002095.article?firstPass=false (accessed on 7 February 2020).

45. Zimmermann, A.; Hofer, S.; Pendl, T.; Kainz, K.; Madeo, F.; Carmona-Gutierrez, D. Yeast as a tool to identify anti-aging compounds. *FEMS Yeast Res.* **2018**, *18*, foy020. [CrossRef] [PubMed]

46. Araújo, K.C.; De M B Costa, E.M.; Pazini, F.; Valadares, M.C.; De Oliveira, V. Bioconversion of quercetin and rutin and the cytotoxicity activities of the transformed products. *Food Chem. Toxicol.* **2013**, *51*, 93–96. [CrossRef] [PubMed]

47. Tajik, N.; Tajik, M.; Mack, I.; Enck, P. The potential effects of chlorogenic acid, the main phenolic components in coffee, on health: A comprehensive review of the literature. *Eur. J. Nutr.* **2017**, *56*, 2215–2244. [CrossRef]

48.  Almeida, A.F.; Borge, G.I.A.; Piskula, M.; Tudose, A.; Tudoreanu, L.; Valentová, K.; Williamson, G.; Santos, C.N. Bioavailability of Quercetin in Humans with a Focus on Interindividual Variation. *Compr. Rev. Food Sci. Food Saf.* **2018**, *17*, 714–731. [CrossRef]
49.  Del Pino-García, R.; Rivero-Pérez, M.D.; González-SanJosé, M.L.; Croft, K.D.; Muñiz, P. Bioavailability of phenolic compounds and antioxidant effects of wine pomace seasoning after oral administration in rats. *J. Funct. Foods* **2016**, *25*, 486–496. [CrossRef]

 foods

*Article*

# Inhibitory Activities of Polyphenolic Extracts of Bangladeshi Vegetables against α-Amylase, α-Glucosidase, Pancreatic Lipase, Renin, and Angiotensin-Converting Enzyme

**Razia Sultana [1], Adeola M. Alashi [1], Khaleda Islam [2], Md Saifullah [3], C. Emdad Haque [4] and Rotimi E. Aluko [1,5,*]**

[1]   Department of Food and Human Nutritional Sciences, University of Manitoba, Winnipeg, MB R3T 2N2, Canada; sultana1@myumanitoba.ca (R.S.); monisola.alashi@umanitoba.ca (A.M.A.)

[2]   Institute of Nutrition and Food Sciences, University of Dhaka, Nilkhet Rd, Dhaka 1000, Bangladesh; arunadu15@gmail.com

[3]   Natural Resources Management Division, Bangladesh Agricultural Research Council, Dhaka 1215, Bangladesh; m.saif@barc.gov.bd

[4]   Natural Resources Institute, University of Manitoba, Winnipeg, MB R3T 2N2, Canada; CEmdad.Haque@umanitoba.ca

[5]   The Richardson Centre for Functional Foods and Nutraceuticals, University of Manitoba, Winnipeg, MB R3T 2N2, Canada

*   Correspondence: rotimi.aluko@umanitoba.ca; Tel.: +1-204-474-9555

Received: 19 May 2020; Accepted: 10 June 2020; Published: 29 June 2020

**Abstract:** The aim of the study was to determine the in vitro enzyme inhibition activities of aqueous polyphenolic extracts of nine popular Bangladeshi vegetables, namely ash gourd, bitter gourd, brinjal, Indian spinach, kangkong, okra, ridge gourd, snake gourd, and stem amaranth. Polyphenolic glycosides were the major compounds present in the extracts. Inhibition of α-amylase (up to 100% at 1 mg/mL) was stronger than α-glucosidase inhibition (up to 70.78% at 10 mg/mL). The Indian spinach extract was the strongest inhibitor of pancreatic lipase activity ($IC_{50}$ = 276.77 µg/mL), which was significantly better than that of orlistat (381.16 µg/mL), a drug. Ash gourd (76.51%), brinjal (72.48%), and snake gourd (66.82%) extracts were the most effective inhibitors of angiotensin-converting enzyme (ACE), an enzyme whose excessive activities have been associated with hypertension. Brinjal also had a significantly higher renin-inhibitory activity than the other vegetable extracts. We conclude that the vegetable extracts may have the ability to reduce enzyme activities that have been associated with hyperglycemia, hyperlipidemia, and hypertension.

**Keywords:** Bangladesh; vegetables; polyphenols; amylase; glucosidase; renin; angiotensin-converting enzyme; lipase; mass spectrometry

## 1. Introduction

Several physiological disturbances can be attributed to the effect of metabolic syndrome, which is a condition characterized by disease conditions such as obesity, dyslipidemia, hyperglycemia, and hypertension [1]. These metabolic syndrome-related diseased are considered major health concerns worldwide. Obesity is considered as one of the main factors in the pathogenesis of cardiovascular diseases such as hypertension, stroke, type 2 diabetes mellitus (T2DM), and cancer [2,3]. Postprandial hyperglycemia (PPHG), a major feature of T2DM, occurs when the blood glucose level increases after consuming a meal and this is an important factor taken into consideration for diabetes management. PPHG can be controlled by inhibiting the enzymes responsible for causing elevated blood glucose [4].

Several epidemiological and empirical studies have reported that consumption of fruits and vegetables containing polyphenol compounds play an important role in inhibiting carbohydrate-hydrolyzing enzymes such as $\alpha$-amylase and $\alpha$-glucosidase [5]. $\alpha$-amylase (EC 3.2.1.1) is responsible for the hydrolysis of $\alpha$-1,4 glucosidic bonds of starches, which are then converted to oligosaccharides. These oligosaccharides are further converted to maltose, glucose, and limit dextrins, causing blood glucose levels to increase [6]. The enzyme $\alpha$-glucosidase (EC 3.2.1.20), located in the brush border of the small intestine enterocytes, is responsible for carbohydrate breakdown and synthesis. $\alpha$-glucosidase is a carbohydrate-hydrolase that releases monosaccharides, and is the main enzyme responsible for increasing blood glucose levels after consumption of a meal. Therefore, subsequent to the initial $\alpha$-amylase hydrolysis of dietary carbohydrates, $\alpha$-glucosidase plays an important role in ensuring the release of absorbable monosaccharides in the intestinal tract [7]. To control PPHG, acarbose, miglitol, and voglibose are drugs that are used as $\alpha$-glucosidase and $\alpha$-amylase inhibitors. However, these drugs not only are expensive but they also affect the gastrointestinal system negatively, with long-term use of these drugs causing flatulence and diarrhea [8].

Digestion of dietary triglycerides occurs in the small intestine by a key enzyme called pancreatic lipase (PL) to release 2-monoglyceride and fatty acids, which are then absorbed. Hence, inhibition of PL (EC 3.1.1.3) activity is regarded as a major approach for obesity prevention and treatment [9–11]. Although orlistat (a drug) has been used as a PL inhibitor, the negative side effects such as bloating, oily spotting, fecal urgency, steatorrhea, and fecal inconsistency have reduced acceptance [12,13].

Another reason for the development of cardiovascular diseases is hypertension, which causes arteriosclerosis, congestive heart failure, coronary heart disease, end-stage renal diseases, myocardial infarction, and stroke [14]. The renin–angiotensin system (RAS) has been found to be one of the major regulatory mechanisms in blood pressure regulation [15]. Within the RAS, renin (EC 3.4.23.15) acts on angiotensinogen to release angiotensin I, which is then cleaved by angiotensin-converting enzyme (ACE) to produce angiotensin II, a potent vasopressor. Therefore, ACE (EC 3.4.15.1) and/or renin inhibition help to manage hypertension and offer cardiovascular protection by limiting the physiological level of angiotensin II. Just like obesity and diabetes, current clinical management of hypertension involves mainly the use of drugs, which also have negative side effects [15].

Different in vitro and in vivo studies have observed that dietary phenolic compounds have many properties beneficial for maintaining human health [16]. In particular, studies have found that phenolic compounds are helpful for inhibiting lipase, alpha-amylase, alpha-glucosidase, ACE, and renin [17–20]. In fact, polyphenol-rich extracts have been shown to reduce blood pressure in spontaneously hypertensive rats and could serve as potential antihypertensive agents [21,22]. Therefore, this research investigated the enzyme inhibition ability of aqueous extracts of nine popular Bangladesh vegetables to determine their potential utility as agents against obesity (PL, $\alpha$-glucosidase, $\alpha$-amylase), diabetes ($\alpha$-glucosidase, $\alpha$-amylase), and hypertension (ACE, renin). Aqueous extracts were used in order to enhance the practical utilization of the research outcome, since the human gastrointestinal tract is highly hydrophilic and soluble compounds are more likely to be absorbed or interact with enzymes than hydrophobic compounds.

## 2. Materials and Methods

### 2.1. Materials

Fresh vegetables, namely ash gourd (BINA CHALKUMRA 1), bitter gourd (BARI KOROLA 1), brinjal (BARI BEGUN 8), Indian spinach (BARI PUISHAK 1), kangkong (BARI GIMAKOLMI 1), okra (BARI DHEROSH 2), ridge gourd (BARI JHINGA 1), snake gourd (BARI CHICHINGA 1), and stem amaranth (BARI DATA 1), were collected from the Bangladesh Agricultural Research Institute (BARI), Gazipur, Bangladesh. The vegetables were dried in the oven at 40 °C, ground into powders, and then stored at −20 °C. Porcine pancreatic $\alpha$-amylase, porcine PL, rat intestinal acetone powder, acarbose, orlistat, 4-nitrophenyl $\alpha$-D-glucopyranoside (PNP), N-(3-[2-furyl]acryloyl)-phenylalanyl glycyl glycine (FAPGG), and rabbit lung ACE were purchased from Sigma-Aldrich (St. Louis, MO, USA). A human recombinant renin inhibitor screening assay kit was purchased from Cayman Chemicals (Ann Arbor, MI, USA). Other analytical grade chemical reagents were obtained from Fisher Scientific Company (Oakville, ON, Canada).

## 2.2. Extraction of Polyphenolic Compounds

Water-soluble free (unbound) polyphenols were extracted using a previous method [23] with minor modifications as follows. Briefly, dried vegetable powders were weighed and one part mixed with 20 parts of distilled water (1:20) in a 500 mL beaker, which was then adjusted to 60 °C and stirred for 2 h [23]. After cooling to ambient temperature, the mixture was centrifuged for 30 min at $5600\times g$, and the supernatant passed through a muslin cloth to collect filtrate #1. The residue was mixed with water (1:20) and the extraction, centrifugation, and filtration processes repeated to collect filtrate #2, which was then combined with filtrate #1. The combined filtrate was partially evaporated in a rotary evaporator (Heidolph Instruments GmbH & CO., Schwabach, Germany) under vacuum (~24 mm Hg) at 60 °C, and the aqueous residue freeze-dried and stored at −20 °C.

## 2.3. UHPLC MS/MS Analysis of Polyphenolic Compounds

An Agilent 1290 UHPLC system (Santa Clara, CA, USA) coupled with an HSS T3 $2.1 \times 100$ mm 1.7 μm column from Waters Corp (Milford, MA, USA) was used to perform UHPLC analysis of the vegetable extracts. Samples were mixed with distilled water, vortexed, and passed through a 0.2 μm filter. Then, a 5 μL portion of the filtrate was injected onto the column followed by elution at a flow rate of 0.5 mL/min using mobile phases A and B (0.1% formic acid in water and 0.1% formic acid in acetonitrile, respectively) at 40 °C. The following gradients were used: initial holding time 0.5 min, mobile phase B ramped up to 50% after 5 min, 95% after 6 min, held for 1 min, and re-equilibrated for 1.5 min. Compounds were identified using a diode array detector at a wavelength range of 230–640 nm in 2 nm increments and a frequency of 5 Hz. The mass spec was carried out in an Agilent 6550 QTOF (Santa Clara, CA, USA) at 200 °C, using a drying gas pressure of 18 psi, 40 psi nebulizer and 350 °C sheath gas, a pressure of 12 psi, and 3500 V capillary with a 1000 V nozzle, and ran in positive ion electrospray at a frequency of 3Hz and acquisition from 30–1700 *m/z*. The MS/MS was performed at a narrow quadruple setting (1.3 atomic mass units), using 10, 20, and 40 eV collision energy and 30–1700 *m/z*. The compounds were identified using MS/MS fragmentation patterns and quantified based on the MS peak area.

## 2.4. Inhibition of α-Amylase Activity

α-amylase inhibition assay was carried out using a previous method [14] with slight modifications. Plant extracts were dissolved in 1 mL of 0.2 mM sodium phosphate buffer, pH 6.9 containing 6 mM NaCl. Then, 100 μL of sample aliquot (0.03–10 mg/mL final concentration) and 100 μL of α-amylase solution were added together in a test tube and incubated for 10 min at 25 °C. A 100 μL amount of 1% starch (previously dissolved in the same buffer, heated, and cooled) was added to the mixture and incubated again at 25 °C for 10 min. Then, 200 μL of 96 mM dinitrosalicylic acid (DNSA), prepared in 2 M sodium potassium tartrate tetrahydrate, were added to terminate the reaction and heated in a water bath at 100 °C for 15 min. Subsequently, a 3 mL amount of double-distilled water was added after the reaction mixture was cooled down to room temperature. From this reaction mixture, a 200 μL aliquot was transferred to a 96-well microplate and absorbance read at 540 nm using a Synergy™ H4 microplate reader (Biotek™, Winooski, VT, USA) at 25 °C. The phosphate buffer was used as a blank and its absorbance subtracted from each well to calculate enzyme activity. Acarbose, a known α-amylase inhibitor, was used as standard and assayed concomitantly with the samples. The inhibitory activity of α-amylase was calculated using the following equation:

$$\text{Inhibition (\%)} = (Ac - As/Ac) \times 100$$

where Ac = Absorbance of the control (no inhibitor) and As = Absorbance of the sample.

*2.5. Inhibition of α-Glucosidase Activity*

α-glucosidase inhibitory activity of the samples was determined according to a previously described method [24] with the following modifications. First, a 300 mg portion of rat intestinal powder was mixed with 9 mL of 0.9% (*w/v*) NaCl solution and centrifuged at 5600× *g* for 30 min, and the supernatant was used as the source of α-glucosidase activity. Plant extracts were dissolved (final concentration of 0.03–10 mg/mL) in 0.1 M sodium phosphate buffer pH 6.9 and 50 µL mixed with 50 µL of the α-glucosidase solution in a 96-well microplate followed by incubation for 10 min at 37 °C. Then, 100 µL 5 mM (PNP) also dissolved in the phosphate buffer were added to each well and the absorbance read at 405 nm in 30 s intervals for 30 min using the Synergy™ H4 microplate reader with temperature maintained at 37 °C. A blank measurement was taken without the addition of the enzyme, and its absorbance was subtracted from each well. Acarbose, an α-glucosidase inhibitor, was used as standard and assayed using the same protocol. The following equation was used to determine the α-glucosidase inhibitory activity of the samples:

$$\text{Inhibition } (\%) = (\text{Ac} - \text{As}) / (\text{Ac}) \times 100$$

where Ac = Absorbance of the control (no inhibitor) and As = Absorbance of the sample.

*2.6. Inhibition of Lipase Activity*

The method described by Tang et al. [25] with slight modifications was used to determine PL inhibition by measuring the release of 4-methyl umbelliferone (4MU) from 4-methyl umbelliferyl oleate (4MUO). PL solution (final concentration of 3.125 U/mL) was prepared in 13 mM Tris-HCl buffer, pH 8.0 containing 1.3 mM CaCl$_2$ and 25 µL added to the mixture containing 225 µL of a 0.5 mM 4-MUO solution and 25 µL of sample (different concentrations) to start the enzyme reaction, followed by incubation for 1 h at 37 °C. The Synergy™ H4 microplate reader was set at 400 nm and used to measure the amount of 4MU released during the reaction. Orlistat, a commonly used pharmacological agent against PL, was used as the standard.

The following equation was used to calculate PL inhibition:

$$\text{Inhibition } (\%) = (\text{Ac} - \text{As}) / (\text{Ac}) \times 100$$

where Ac = Absorbance of control (no inhibitor) and As = Absorbance of the sample.

*2.7. ACE Inhibition Assay*

The method described by Alashi et al. [21] was used to determine ACE-inhibitory activity of the polyphenolic extracts. ACE, FAPGG (ACE substrate), and samples were individually dissolved in 50 mM Tris-HCl buffer, pH 7.5 containing 0.3 M NaCl. A 10 µL aliquot of ACE (final reaction activity 25 mU) was added to each well containing 170 µL of 0.5 mM FAPPG and 20 µL of samples at 37 °C. The buffer was used as blank (uninhibited reaction), while captopril, an ACE-inhibitory drug, was used as standard and assayed using a similar protocol. Absorption was read at 345 nm at 1 min intervals for 30 min to determine the reaction rate. The slope of the blank or sample reactions was used to calculate the percentage ACE inhibition as follows:

$$\text{ACE inhibition } (\%) = \left( \frac{\text{Slope } (A/\text{min})_{\text{blank}} - \text{Slope } (A/\text{min})_{\text{sample}}}{\text{Slope } (A/\text{min})_{\text{blank}}} \right) \times 100)$$

### 2.8. Renin Inhibition Assay

The renin-inhibitory activity of samples was determined using a previously described method [26]. Briefly, 20 µL of the substrate, 160 µL of assay buffer, and 10 µL of distilled water were mixed and added to the background well. Then, 20 µL of the substrate, 150 µL of assay buffer, and 10 µL of water were mixed into the control (uninhibited) wells, whereas the sample (inhibited) wells contained same reagents except that the water was replaced with 10 µL of samples (0.5 mg/mL assay concentration). This was followed by the addition of 10 µL of renin solution (dissolved in the assay buffer) to the control and sample wells to initiate enzyme reaction; the microplate was shaken for 10 s to ensure adequate mixing of the reagents and then incubated at 37 °C for 10 min in the dark. Enzyme catalytic activity was measured as the fluorescence intensity (FI) measured at excitation and emission wavelengths of 340 and 490 nm, respectively. Enzyme inhibition was calculated as follows after subtracting the FI of the background well from the control and sample wells:

$$\text{Renin inhibition } (\%) = (\text{FI of control well} - \text{FI of sample well}) / (\text{FI of control well}) \times 100$$

### 2.9. Statistical Analysis

A minimum of duplicate assays were used to find out the mean values and standard deviations. For statistical analysis, analysis of variance (Kruskal–Wallis ANOVA) was used, while significant differences ($p < 0.05$) between mean values were determined by the Duncan's multiple range tests. The IBM SPSS statistical package (version 24, Armonk, NY, USA) was used for all statistical analyses.

## 3. Results

### 3.1. UHPLC MS/MS Analysis

The major polyphenolic compounds identified in the vegetable extracts are shown in Table 1. More polyphenolic compounds were detected from Indian spinach, kangkong, and okra as compared to other vegetable extracts. The identified compounds were mainly chlorogenic acid and the glycosides of quercetin, vitexin hexoside, and kaempferol. Kaempferol O-sophoroside was particularly present at a very high concentration in snake gourd and may be the main compound that determined the bioactive properties of this extract.

**Table 1.** Main polyphenolic compounds of aqueous extracts of vegetables *.

| Samples | Polyphenolic Compounds | Retention Time (min) | *m/z* (Da) | MS/MS (Da) | Concentration (mg/g) A [1] | Concentration (mg/g) B [2] |
|---|---|---|---|---|---|---|
| Ash Gourd | Vitexin ramnoside | 3.2 | 579 | 433, 313, 283 | 1.23 | 0.77 |
| Brinjal | Chlorogenic acid | 2.4 | 355 | 163 | 5.63 | 4.07 |
| Indian spinach | Vitexin arabinoside | 3.0 | 465 | 433, 313, 283 | 8.96 | 3.40 |
| | Vitexin | 3.2 | 433 | 313, 283 | 2.82 | 1.07 |
| | Kaempferol O-rutinoside | 3.4 | 595 | 449, 287 | 0.12 | 0.05 |
| Kangkong | Dicaffeoyl quinic acid | 3.5 | 517 | 163 | 6.92 | 3.72 |
| | Quercetin hexoside | 3.4 | 465 | 163 | 4.46 | 2.40 |
| | Chlorogenic, isochlorogenic acid | 2.5 | 355 | 163 | 10.12 | 5.43 |
| Okra | Quercetin O-sophoroside | 3.0 | 627 | 301 | 3.33 | 2.28 |
| | Quercetin O-hexoside | 3.4 | 465 | 301 | 2.63 | 1.80 |
| | Quercetin malonyl O-hexoside | 3.5 | 551 | 301 | 0.66 | 0.45 |
| Snake gourd | Kaempferol O-sophoroside | 3.1 | 611 | 287 | 81.35 | 57.15 |

* Compounds in bitter gourd, ridge gourd, and stem amaranth could not be identified. [1] Dried polyphenolic extract. [2] Dried vegetable powder.

### 3.2. α-Amylase Inhibition

The inhibitory activity of α-amylase obtained in this study by the phenolic extracts was mostly dose-dependent, although at higher concentrations than acarbose, the standard compound (Figure 1). However, for brinjal and stem amaranth, decreases at sample concentrations >0.6 and 0.8 mg/mL, respectively, were observed. Only ash gourd had detectable activity at 0.2 mg/mL, whereas only brinjal, Indian spinach, and snake gourd had high inhibition levels at 0.4 mg/mL. Kangkong had no detectable activity at 0.2–0.6 mg/mL. Among the vegetable extracts, the ridge gourd achieved 100% inhibition at 1 mg/mL concentration, which is significantly ($p < 0.05$) better than the other extracts. The okra (40.82% ± 1.88%) and stem amaranth (36.84% ± 0.49%) showed the lowest levels of α-amylase inhibition, respectively, at 1 mg/mL.

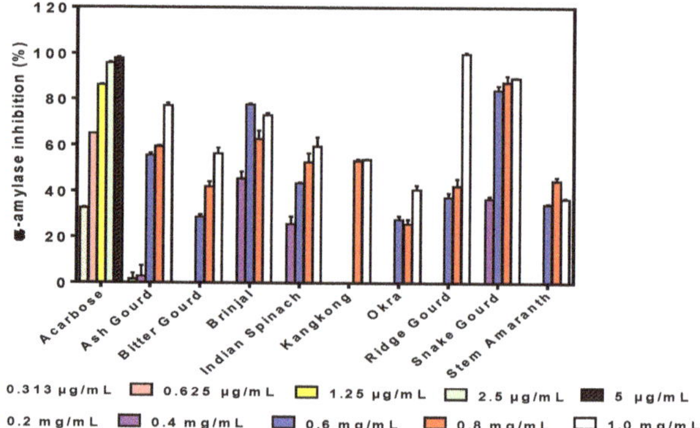

**Figure 1.** Inhibition of α-amylase activity at different concentrations of acarbose and aqueous polyphenolic extracts of vegetables. Bars are means ($n = 3$) ± standard deviation.

### 3.3. α-Glucosidase Inhibition

A mostly dose-dependent effect was also observed as the inhibitory activity of α-glucosidase increased with increasing concentration for almost all the samples (Figure 2). However, decreases in α-glucosidase inhibition were observed at sample concentrations >6 and 8 mg/mL for bitter gourd and stem amaranth, respectively. The highest inhibition was observed for brinjal (70.78% ± 3.45%) at 10 mg/mL, which is similar to that of acarbose (69.36% ± 0.80% at 0.5 mg/mL), a purified synthetic inhibitor of α-glucosidase. With the exception of brinjal (higher activity) and stem amaranth (lower activity), the inhibitory values obtained for the vegetable extracts were similar at 10 mg/mL, but snake gourd had the lowest inhibition at 0.2 mg/mL.

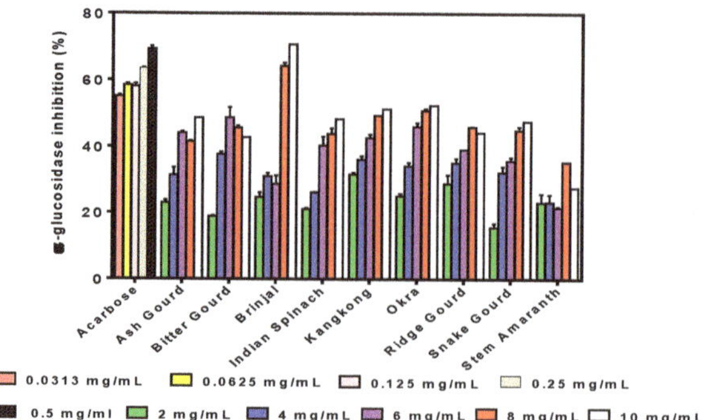

**Figure 2.** Inhibition of α-glucosidase activity at different concentrations of acarbose and aqueous polyphenolic extracts of vegetables. Bars are means ($n = 3$) ± standard deviation.

### 3.4. Pancreatic Lipase (PL)-Inhibitory Activity

PL inhibition was strong for all the vegetable extracts, which enabled $IC_{50}$ calculation as shown in Figure 3. Because lower $IC_{50}$ values indicate stronger inhibitory activity, the results obtained show that Indian spinach (276.77 ± 4.95 µg/mL) was the most active with a significantly ($p < 0.05$) lower value than orlistat. Brinjal (397.22 ± 2.36 µg/mL), okra (427.94 ± 1.40 µg/mL), and kangkong (482.04 ± 0.67 µg/mL) also had strong activities, although lower than orlistat (381.16 µg/mL). It is also noticeable that the stem amaranth extract had the weakest PL inhibition ($IC_{50}$ = 723.394 ± 2.36 µg/mL), just as it showed the lowest inhibitions of α-amylase and α-glucosidase inhibitory activities.

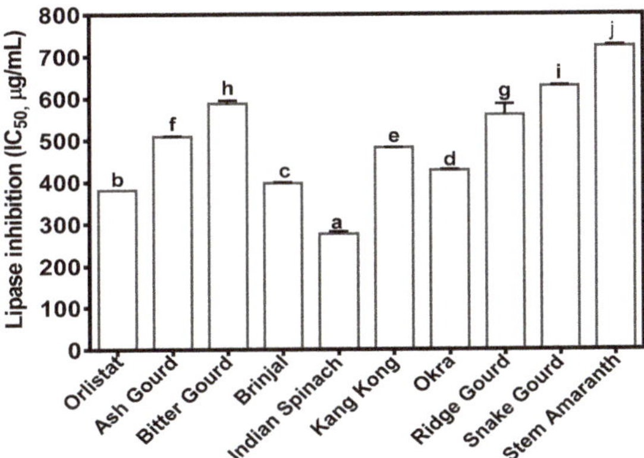

**Figure 3.** $IC_{50}$ values ($n = 3$ ± standard deviation) for the inhibition of pancreatic lipase (PL) activity by orlistat and aqueous polyphenolic extracts of vegetables. Bars with different letters have mean values that are significantly ($p < 0.05$) different.

### 3.5. Angiotensin-Converting Enzyme (ACE)-Inhibitory Activity

Different levels of ACE inhibition were obtained for the vegetable extracts but, in general, the ash gourd (76.51% ± 0.25%), brinjal (72.48% ± 0.02%), and snake gourd (66.82% ± 0.99%) were the most active (Figure 4). All the vegetable extracts had significantly ($p < 0.05$) weaker inhibition than captopril, the ACE-inhibitory drug. It is also observable that ash gourd also showed a good record for inhibiting α-amylase and α-glucosidase. Stem amaranth showed the lowest activity for all the concentrations and the lowest ACE inhibition, which is consistent with the observed poor inhibitory activities against PL, amylase, and glucosidase.

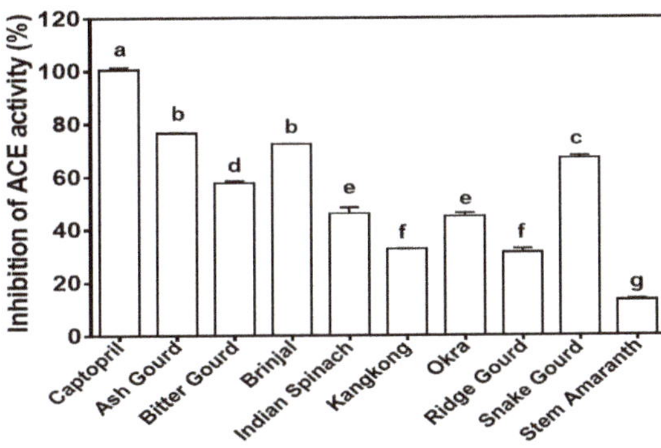

**Figure 4.** Inhibition of angiotensin-converting enzyme (ACE) activity ($n = 3$ ± standard deviation) by 1 mg/mL captopril and aqueous polyphenolic extracts of vegetables. Bars with different letters have mean values that are significantly ($p < 0.05$) different.

### 3.6. Renin-Inhibitory Activity

The highest renin inhibitory activity was exhibited by brinjal (79.64% ± 7.63%) at 0.5 mg/mL as compared to 99.11% ± 1.75% for aliskiren (drug) at 0.05 mg/mL (Figure 5). The snake gourd had the significantly lowest renin inhibition, which was not significantly ($p > 0.05$) different from the stem amaranth with no detectable renin inhibition.

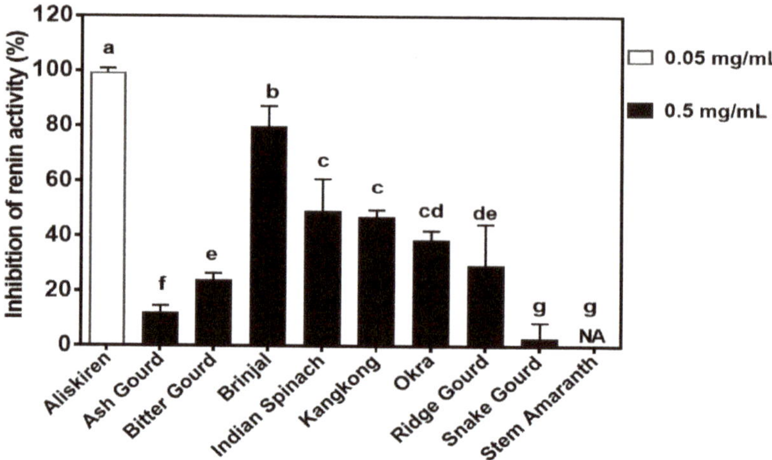


**Figure 5.** Inhibition of renin activity ($n = 3 \pm$ standard deviation) by aliskiren and aqueous polyphenolic extracts of vegetables. Bars with different letters have mean values that are significantly ($p < 0.05$) different.

## 4. Discussion

UHPLC MS/MS analysis is considered an effective analysis method because it provides high selectivity, high sensitivity, and the potentiality for robust and accurate analysis identification of compounds [27]. As shown in Table 1, the dominant phenolic compounds were the glycosides, which reflect the aqueous extraction. The water solubility properties of the phenolic compounds could enhance their interactions with target enzymes within the hydrophilic environment of the gastrointestinal tract. For example, the aqueous extracts of some bean varieties were shown to have stronger lipase inhibition than the ethanolic extracts [28]. Inhibition of $\alpha$-amylase was mainly dose-dependent except for brinjal and stem amaranth, where their decreases occurred at higher sample concentrations. However, the dose-dependent inhibitory activity pattern observed for most of the samples was also reported for $\alpha$-amylase inhibition by bitter gourd extract [29]. The decreased $\alpha$-amylase inhibition at high concentrations of brinjal and stem amaranth may be due to increased interactions between the polyphenol molecules, which reduced interactions with the enzyme protein. The weaker $\alpha$-amylase inhibitory activity when compared to acarbose, which is an approved drug, suggests lower enzyme binding intensity by the polyphenolic compounds. A previous study also reported that the ethanolic extract of bitter gourd exhibited lower $\alpha$-amylase activity than acarbose [29]. The high inhibitory activity of snake gourd (89.26% ± 0.23%) may be attributed to the kaempferol O-sophoroside, although the contribution of other non-identified compounds cannot be discounted. The low inhibitory activity of okra suggests that the quercetin glycosides, which were the main identified compounds in the extract, are not very effective inhibitors of $\alpha$-amylase. In contrast, other extracts that contained vitexin, chlorogenic acid, and dicaffeoyl quinic acid had higher inhibitions of $\alpha$-amylase activity when compared to okra that contained mainly quercetin glycosides. $\alpha$-amylase inhibitors are also referred to as starch blockers because they prevent starch absorption. Digestive amylase enzyme and other secondary enzymes play a critical role in breaking down complex carbohydrates such as starch, without which they cannot be absorbed because polysaccharides need to be broken down first into monosaccharides for absorption to take place [30]. The association between α-amylase inhibition and its potential contribution to the management of T2DM with phenolic extracts have been investigated for other vegetables [31,32].

α-glucosidase is another enzyme that makes glucose available in the body by breaking down oligo- and disaccharides, converting them into absorbable monosaccharides. Therefore, in order to reduce serum glucose levels and manage related diseases, α-glucosidase inhibition has been proposed as a suitable approach [33,34]. This work showed that the vegetable extracts had weaker α-glucosidase inhibition than acarbose, the drug. In fact, only brinjal extract at 8 and 10 mg/mL produced similar inhibitory values to acarbose at 0.25 mg/mL. In contrast, a nanoparticulated ethanolic extract of bitter gourd was reported to have stronger α-glucosidase inhibition than acarbose [29]. Nwanna, Ibukun, and Oboh [35] reported 50% inhibition of α-glucosidase activity for a similar eggplant variety using 63.24 ± 0.30 µg/mL, which is stronger than the results obtained in this work. The difference may be due to their use of solvent extraction, which could have isolated more compounds than the aqueous extraction used in the present work. The observed variations in identifiable polyphenolic compounds of the vegetable extracts did not have strong effects in influencing α-glucosidase inhibition. However, the results obtained in this work agree with a previous work [30], which suggested that brinjal phenolics may have potential for use in controlling T2DM because of the strong glucosidase-inhibitory property. A previous work [36] reported 66.64% (at 2.5 mg/mL), which is higher than the 48.65% (at 6 mg/mL) obtained in this work. The α-glucosidase inhibitory properties of a sample have been shown to be dependent on the type of extracting solvent; for example, ethyl acetate extract had stronger activity than hexane, methanol, and chloroform extracts of bitter gourd [36]. Therefore, the weaker activity obtained for the bitter gourd extract in this work may be due to the aqueous extraction as compared to the ethyl acetate extract.

Lipase is secreted in the pancreas and is used as a catalyst to aid triglyceride hydrolysis in the stomach, and this hydrolysis is completed by intestinal lipase in the small intestine [28,37]. It has been estimated that about 50–70% of total dietary fats are hydrolyzed for absorption by PL [38], which emphasizes the importance of this enzyme in calorie release from diets. Surprisingly, orlistat is the only FDA-approved drug for inhibiting PL. It was reported that orlistat can prevent approximately 30% of dietary fat absorption; however, regular usage is associated with some undesirable side effects like flatulence, diarrhea, oily spotting, incontinence, abdominal cramping, and fecal urgency [39]. Therefore, the search for active PL inhibitors, especially natural compounds with potentially fewer side effects, has become important. Results from this work confirmed the PL-inhibitory activity of the vegetable extracts, especially the stronger effect of Indian spinach as compared to orlistat. Therefore, the Indian spinach extract could serve as a potential agent for use in limiting PL-dependent digestion of dietary lipids. While there may be contributions from other polyphenolic compounds, the results suggest that the unique combination of vitexin with the kaempferol and vitexin glycosides contributed to the stronger PL inhibition by Indian spinach. This is because ash gourd and snake gourd, which contain vitexin and kaempferol glycosides, respectively, but not both types of polyphenols, were not as active as Indian spinach. The activities obtained in this work are weaker than the 92.0 ± 6.3 to 128.5 ± 7.4 µg/mL reported for the polyphenolic-rich extracts of common bean varieties [10]. The differences may be due to the type of sample and the extraction solvent since acetone was used as compared to aqueous extraction in the present work.

ACE is responsible for the formation of angiotensin II, a powerful vasoconstrictor; excessive physiological levels of angiotensin II lead to hypertension [15]. Therefore, ACE activity inhibition has been used to prevent and treat various diseases like heart failure, myocardial infarction, nephropathy, and even diabetes. However, apart from the different negative side effects associated with the use of pharmacological agents, ACE-inhibitory drugs are not permitted for use during pregnancy due to potential damage to the fetus [40]. Therefore, foods may serve as sources of ACE-inhibitory compounds with less harmful effects. For example, flavanols have been shown to inhibit in vitro activity of ACE as well as in isolated rat kidney membranes [41]. The vegetable extracts had lower ACE inhibition than captopril, which is consistent with the strong antihypertensive activity associated with standard drugs. However, the advantage of natural products is the reduced risk of negative side effects that could make them preferred antihypertensive agents when compared to synthetic drugs. Moreover, it is possible to incorporate these vegetable extracts into foods for regular consumption to combat the incidence of hypertension [21]. The results suggest that the presence of multiple compounds as identified in this work produced no strong synergistic effect since the strongest ACE inhibition was associated with samples where one polyphenolic compound predominate. However, it should be emphasized that other non-identified compounds may have contributed to the observed activities. The results are consistent with previous reports, which have reported that ACE-inhibitory activity varies because of differences in polyphenolic compounds of different plant extracts [21,42–45].

Renin inhibition has also been shown to be an effective means of reducing blood pressure because it catalyzes the rate-determining step (conversion of angiotensinogen to angiotensin I) in the RAS pathway [15,46]. However, effective renin-inhibitory drugs are rare and only one (aliskiren) has so far been approved as an antihypertensive agent [47]. Therefore, natural compounds that inhibit renin activity could enhance the management of hypertension. This is possible because polyphenolic compounds such as saponins have been shown to bind renin protein [48] and produce blood pressure-lowering effects after oral administration to rats [49]. As shown in Figure 5, the vegetable extracts had weaker renin inhibition than aliskiren even though the drug was used at 10 times lower the concentration of the extracts. The weaker renin inhibition by the vegetable extracts is consistent with the high purity and chemical potency of the synthesized drug. However, brinjal extract had the strongest renin inhibition among the vegetables, which suggests that this sample could serve as a potential natural product to reduce renin activity. Unlike ACE inhibition, the presence of multiple polyphenolic compounds may have worked synergistically to contribute to the renin-inhibitory activity of Indian spinach, kangkong, and okra. Reports of polyphenol-dependent inhibition of renin activity are scant, but a previous work with aqueous vegetable leaf extracts reported <40% inhibitions, although at a lower concentration of 0.25 mg/mL [22]. Strong inhibition of in vitro renin activity was also reported for purified soybean saponins with an $IC_{50}$ value of 59.9 µg/mL [49]. The high purity of the saponin preparation could have contributed to the high renin inhibition when compared to the mixture of polyphenolic compounds used in this work.

## 5. Conclusions

Results from this work have shown that the aqueous extracts of vegetables could inhibit enzyme activities with respect to obesity, diabetes, and hypertension. The extracts were more effective inhibitors of $\alpha$-amylase than $\alpha$-glucosidase, which suggests a strong potential to limit excessive glucose release during digestion because starch hydrolysis is the rate-limiting step. Ability of the extracts to also inhibit PL suggests potential use as agents to limit the release of fatty acids during intestinal digestion, which could assist in reducing hyperlipidemia. Inhibition of ACE and renin activities are indications of potential blood pressure-reducing ability, which could also make the extracts function as antihypertensive agents. Overall, Indian spinach and brinjal extracts produced the most promising inhibitory effects on the five enzymes studied in this work, whereas stem amaranth extract was the poorest. However, in vivo determination of enzyme inhibitory activities using suitable animal disease models is required to confirm potential health benefits.

**Author Contributions:** Conceptualization, funding acquisition, project administration, R.E.A., C.E.H., K.I., and M.S.; methodology, R.E.A., A.M.A., and R.S.; resources, R.E.A. and C.E.H.; writing—original draft preparation, R.S.; writing—review and editing, supervision, formal analysis, R.E.A. and A.M.A. All authors have read and agreed to the published version of the manuscript.

**Funding:** This research was funded by an operating grant from the International Development Research Centre (IDRC) and Global Affairs Canada through the Canadian International Food Security Research Fund (CIFSRF) Project 108163-002 on reducing dietary-related risks associated with non-communicable diseases in Bangladesh.

**Conflicts of Interest:** The authors declare no conflict of interest. The funders had no role in the design of the study; in the collection, analyses, or interpretation of data; in the writing of the manuscript; or in the decision to publish the results.

## References


1. Miglani, N.; Bains, K. Interplay between proteins and metabolic syndrome—A review. *Crit. Rev. Food Sci. Nutr.* **2017**, *57*, 2483–2496. [CrossRef]
2. Handy, D.E.; Castro, R.; Loscalzo, J. Epigenetic modifications: Basic mechanisms and role in cardiovascular disease. *Circulation* **2011**, *123*, 2145–2156. [CrossRef]
3. Tucci, S.A.; Boyland, E.J.; Halford, J.C. The role of lipid and carbohydrate digestive enzyme inhibitors in the management of obesity: A review of current and emerging therapeutic agents. *Diabetes Metab. Syndr. Obes.* **2010**, *3*, 125–143. [CrossRef]
4. Poovitha, S.; Parani, M. In vitro and in vivo α-amylase and α-glucosidase inhibiting activities of the protein extracts from two varieties of bitter gourd (*Momordica charantia* L.). *BMC Complement. Altern. Med.* **2016**, *16*, 185. [CrossRef] [PubMed]
5. Wongsa, P.; Chaiwarit, J.; Zamaludien, A. In vitro screening of phenolic compounds, potential inhibition against α-amylase and α-glucosidase of culinary herbs in Thailand. *Food Chem.* **2012**, *131*, 964–971. [CrossRef]
6. Lee, S.Y.; Kim, S.; Sweet, R.M.; Suh, S.W. Crystallization and a preliminary X-ray crystallographic study of α-amylase from *Bacillus licheniformis*. *Arch. Biochem. Biophys.* **1991**, *291*, 255–257. [CrossRef]
7. Patil, P.; Mandal, S.; Tomar, S.K.; Anand, S. Food protein-derived bioactive peptides in management of type 2 diabetes. *Eur. J. Nutr.* **2015**, *54*, 863–880. [CrossRef] [PubMed]
8. Thanakosai, W.; Phuwapraisirisan, P. First identification of α-glucosidase inhibitors from okra (*Abelmoschus esculentus*) seeds. *Nat. Prod. Commun.* **2013**, *8*, 1085–1088. [CrossRef] [PubMed]
9. De La Garza, A.L.; Milagro, F.I.; Boque, N.; Campión, J.; Martínez, J.A. Natural inhibitors of pancreatic lipase as new players in obesity treatment. *Planta Med.* **2011**, *77*, 773–785. [CrossRef]
10. Ombra, M.N.; d'Acierno, A.; Nazzaro, F.; Spigno, P.; Riccardi, R.; Zaccardelli, M.; Fratianni, F. Alpha-amylase, α-glucosidase and lipase inhibiting activities of polyphenol-rich extracts from six common bean cultivars of Southern Italy, before and after cooking. *Int. J. Food Sci. Nutr.* **2018**, *69*, 824–834. [CrossRef]
11. Yoshikawa, M.; Shimoda, H.; Nishida, N.; Takada, M.; Matsuda, H. *Salacia reticulata* and its polyphenolic constituents with lipase inhibitory and lipolytic activities have mild antiobesity effects in rats. *J. Nutr.* **2002**, *132*, 1819–1824. [CrossRef] [PubMed]
12. Ballinger, A.; Peikin, S.R. Orlistat: Its current status as an anti-obesity drug. *Eur. J. Pharmacol.* **2002**, *440*, 109–117. [CrossRef]
13. Drew, B.S.; Dixon, A.F.; Dixon, J.B. Obesity management: Update on orlistat. *Vasc. Health Risk Manag.* **2007**, *3*, 817–821. [PubMed]
14. Chiou, S.Y.; Sung, J.M.; Huang, P.W.; Lin, S.D. Antioxidant, antidiabetic, and antihypertensive properties of *Echinacea purpurea* flower extract and caffeic acid derivatives using in vitro models. *J. Med. Food* **2017**, *20*, 171–179. [CrossRef] [PubMed]
15. Aluko, R.E. Food protein-derived renin-inhibitory peptides: In vitro and in vivo properties. *J. Food Biochem.* **2019**, *43*, e12648. [CrossRef]
16. Da-Costa-Rocha, I.; Bonnlaender, B.; Sievers, H.; Pischel, I.; Heinrich, M. *Hibiscus sabdariffa* L.—A phytochemical and pharmacological review. *Food Chem.* **2014**, *165*, 424–443. [CrossRef]
17. Ademiluyi, A.O.; Oboh, G.; Ogunsuyi, O.B.; Oloruntoba, F.M. A comparative study on antihypertensive and antioxidant properties of phenolic extracts from fruit and leaf of some guava (*Psidium guajava* L.) varieties. *Comp. Clin. Pathol.* **2016**, *25*, 363–374. [CrossRef]
18. Chiang, Y.C.; Chen, C.L.; Jeng, T.L.; Sung, J.M. In vitro inhibitory effects of cranberry bean (*Phaseolus vulgaris* L.) extracts on aldose reductase, α-glucosidase and α-amylase. *Int. J. Food Sci. Technol.* **2014**, *49*, 1470–1479. [CrossRef]
19. Chiou, S.Y.; Lai, J.Y.; Liao, J.A.; Sung, J.M.; Lin, S.D. In vitro inhibition of lipase, α-amylase, α-glucosidase, and angiotensin-converting enzyme by defatted rice bran extracts of red-pericarp rice mutant. *Cereal Chem.* **2018**, *95*, 167–176. [CrossRef]

20. Gondoin, A.; Grussu, D.; Stewart, D.; McDougall, G.J. White and green tea polyphenols inhibit pancreatic lipase in vitro. *Food Res. Int.* **2010**, *43*, 1537–1544. [CrossRef]

21. Alashi, A.M.; Taiwo, K.A.; Oyedele, D.; Adebooye, O.C.; Aluko, R.E. Antihypertensive properties of aqueous extracts of vegetable leaf-fortified bread after oral administration to spontaneously hypertensive rats. *Int. J. Food Sci. Technol.* **2018**, *53*, 1705–1716. [CrossRef]

22. Oluwagunwa, O.A.; Alashi, A.M.; Aluko, R.E. *Solanum Macrocarpon* leaf extracts reduced blood pressure and heart rate after oral administration to spontaneously hypertensive rats. *Curr. Top. Nutraceut. Res.* **2019**, *17*, 282–290. [CrossRef]

23. Olarewaju, O.A.; Alashi, A.M.; Aluko, R.E. Antihypertensive effect of aqueous polyphenol extracts of *Amaranthus viridis* and *Telfairia occidentalis* leaves in spontaneously hypertensive rats. *J. Food Bioact.* **2018**, *1*, 166–173. [CrossRef]

24. Ranilla, L.G.; Kwon, Y.I.; Apostolidis, E.; Shetty, K. Phenolic compounds, antioxidant activity and in vitro inhibitory potential against key enzymes relevant for hyperglycemia and hypertension of commonly used medicinal plants, herbs and spices in Latin America. *Bioresour. Technol.* **2010**, *101*, 4676–4689. [CrossRef] [PubMed]

25. Tang, Y.; Zhang, B.; Li, X.; Chen, P.X.; Zhang, H.; Liu, R.; Tsao, R. Bound phenolics of quinoa seeds released by acid, alkaline, and enzymatic treatments and their antioxidant and α-glucosidase and pancreatic lipase inhibitory effects. *J. Agric. Food Chem.* **2016**, *64*, 1712–1719. [CrossRef]

26. Sonklin, C.; Alashi, M.A.; Laohakunjit, N.; Kerdchoechuen, O.; Aluko, R.E. Identification of antihypertensive peptides from mung bean protein hydrolysate and their effects in spontaneously hypertensive rats. *J. Funct. Foods* **2019**, *64*, 103635. [CrossRef]

27. Verdu, C.F.; Gatto, J.; Freuze, I.; Richomme, P.; Laurens, F.; Guilet, D. Comparison of two methods, UHPLC-UV and UHPLC-MS/MS, for the quantification of polyphenols in cider apple juices. *Molecules* **2013**, *18*, 10213–10227. [CrossRef]

28. Vijayaraj, P.; Nakagawa, H.; Yamaki, K. Cyanidin and cyanidin-3-glucoside derived from *Vigna unguiculata* act as noncompetitive inhibitors of pancreatic lipase. *J. Food Biochem.* **2019**, *43*, e12774. [CrossRef]

29. Chizoba, E.F.G.; Suneetha, J.W.; Maheswari, K.; Sucharitha, D.S.; Prasad, T.N.V.K.V. In vitro carbohydrate digestibility of copper nanoparticulated bitter gourd extract. *J. Nutr. Food Sci.* **2016**, *6*, 1000482. [CrossRef]

30. Pinto, M.D.S.; Ranilla, L.G.; Apostolidis, E.; Lajolo, F.M.; Genovese, M.I.; Shetty, K. Evaluation of antihyperglycemia and antihypertension potential of native Peruvian fruits using in vitro models. *J. Med. Food* **2009**, *12*, 278–291. [CrossRef]

31. Cheplick, S.; Kwon, Y.I.; Bhowmik, P.; Shetty, K. Phenolic-linked variation in strawberry cultivars for potential dietary management of hyperglycemia and related complications of hypertension. *Bioresour. Technol.* **2010**, *101*, 404–413. [CrossRef] [PubMed]

32. Oboh, G.; Akinyemi, A.J.; Ademiluyi, A.O. Inhibition of α-amylase and α-glucosidase activities by ethanolic extract of Telfairia occidentalis (fluted pumpkin) leaf. *Asian Pac. J. Trop. Biomed.* **2012**, *2*, 733–738. [CrossRef]

33. Awosika, T.O.; Aluko, R.E. Inhibition of the in vitro activities of α-amylase, α-glucosidase and pancreatic lipase by yellow field pea (*Pisum sativum* L.) protein hydrolysates. *Int. J. Food Sci. Technol.* **2019**, *54*, 2021–2034. [CrossRef]

34. Nanok, K.; Sansenya, S. α-Glucosidase, α-amylase, and tyrosinase inhibitory potential of capsaicin and dihydrocapsaicin. *J. Food Biochem.* **2020**, *44*, e13099. [CrossRef]

35. Nwanna, E.; Ibukun, E.O.; Oboh, G. Inhibitory effects of methanolic extracts of two eggplant species from South-western Nigeria on starch hydrolysing enzymes linked to type-2 diabetes. *Afr. J. Pharm. Pharmacol.* **2013**, *7*, 1575–1584. [CrossRef]

36. Sulaiman, S.F.; Ooi, K.L.; Supriatno. Antioxidant and α-glucosidase inhibitory activities of cucurbit fruit vegetables and identification of active and major constituents from phenolic-rich extracts of *Lagenaria siceraria* and *Sechium edule*. *J. Agric. Food Chem.* **2013**, *61*, 10080–10090. [CrossRef]

37. Patil, M.; Patil, R.; Bhadane, B.; Mohammad, S.; Maheshwari, V. Pancreatic lipase inhibitory activity of phenolic inhibitor from endophytic *Diaporthe arengae*. *Biocat. Agric. Biotechnol.* **2017**, *10*, 234–238. [CrossRef]

38. Embleton, J.K.; Pouton, C.W. Structure and function of gastro-intestinal lipases. *Adv. Drug Deliv. Rev.* **1997**, *25*, 15–32. [CrossRef]

39. Seyedan, A.; Alshawsh, M.A.; Alshagga, M.A.; Koosha, S.; Mohamed, Z. Medicinal plants and their inhibitory activities against pancreatic lipase: A review. *Evid. Based Complement. Altern. Med.* **2015**, *2015*, 973143. [CrossRef]

40. Cooper, W.O.; Hernandez-Diaz, S.; Arbogast, P.G.; Dudley, J.A.; Dyer, S.; Gideon, P.S.; Ray, W.A. Major congenital malformations after first-trimester exposure to ACE inhibitors. *N. Engl. J. Med.* **2006**, *354*, 2443–2451. [CrossRef]

41. Actis-Goretta, L.; Ottaviani, J.I.; Fraga, C.G. Inhibition of angiotensin converting enzyme activity by flavanol-rich foods. *J. Agric. Food Chem.* **2006**, *54*, 229–234. [CrossRef] [PubMed]

42. Afonso, J.; Passos, C.P.; Coimbra, M.A.; Silva, C.M.; Soares-da-Silva, P. Inhibitory effect of phenolic compounds from grape seeds (*Vitis vinifera* L.) on the activity of angiotensin I converting enzyme. *LWT Food Sci. Technol.* **2013**, *54*, 265–270. [CrossRef]

43. Al Shukor, N.; Van Camp, J.; Gonzales, G.B.; Staljanssens, D.; Struijs, K.; Zotti, M.J.; Smagghe, G. Angiotensin-converting enzyme inhibitory effects by plant phenolic compounds: A study of structure activity relationships. *J. Agric. Food Chem.* **2013**, *61*, 11832–11839. [CrossRef] [PubMed]

44. Guerrero, L.; Castillo, J.; Quiñones, M.; Garcia-Vallvé, S.; Arola, L.; Pujadas, G.; Muguerza, B. Inhibition of angiotensin-converting enzyme activity by flavonoids: Structure-activity relationship studies. *PLoS ONE* **2012**, *7*, e49493. [CrossRef]

45. Ojeda, D.; Jiménez-Ferrer, E.; Zamilpa, A.; Herrera-Arellano, A.; Tortoriello, J.; Alvarez, L. Inhibition of angiotensin convertin enzyme (ACE) activity by the anthocyanins delphinidin- and cyanidin-3-O-sambubiosides from *Hibiscus sabdariffa*. *J. Ethnopharmacol.* **2010**, *127*, 7–10. [CrossRef]

46. Pechanova, O.; Barta, A.; Koneracka, M.; Zavisova, V.; Kubovcikova, M.; Klimentova, J.; Cebova, M. Protective effects of nanoparticle-loaded aliskiren on cardiovascular system in spontaneously hypertensive rats. *Molecules* **2019**, *24*, 15. [CrossRef]

47. Wood, J.M.; Maibaum, J.; Rahuel, J.; Grutter, M.G.; Cohen, N.-C.; Rasetti, V.; Bedigian, M.P. Structure-based design of aliskiren, a novel orally effective renin inhibitor. *Biochem. Biophys. Res. Commun.* **2003**, *308*, 698–705. [CrossRef]

48. Tavassoli, Z.; Taghdir, M.; Ranjbar, B. Renin inhibition by soyasaponin I: A potent native anti-hypertensive compound. *J. Biomol. Struct. Dyn.* **2018**, *36*, 166–176. [CrossRef]

49. Hiwatashi, K.; Shirakawa, H.; Hori, K.; Yoshiki, Y.; Suzuki, N.; Hokari, M.; Komai, M.; Takahashi, S. Reduction of blood pressure by soybean saponins, renin inhibitors from soybean, in spontaneously hypertensive rats. *Biosci. Biotechnol. Biochem.* **2010**, *74*, 2310–2312. [CrossRef]

*Article*

# Improvement of Fatty Acid Profile in Durum Wheat Breads Supplemented with *Portulaca oleracea* L. Quality Traits of Purslane-Fortified Bread

**Maria Grazia Melilli** [1,*], **Vita Di Stefano** [2,*], **Fabiola Sciacca** [3], **Antonella Pagliaro** [1], **Rosaria Bognanni** [1], **Salvatore Scandurra** [1], **Nino Virzì** [3], **Carla Gentile** [2] **and Massimo Palumbo** [3]

[1] Institute for Agricultural and Forest Systems in the Mediterranean, National Council of Research, 95128 Catania, Italy; antonella.pagliaro@isafom.cnr.it (A.P.); rosaria.bognanni@cnr.it (R.B.); salvatore.scandurra@cnr.it (S.S.)

[2] Department of Biological, Chemical, and Pharmaceutical Science and Technology (STEBICEF), University of Palermo, 90123 Palermo, Italy; carla.gentile@unipa.it

[3] CREA Council for Agricultural Research and Economics—Research Centre for Cereal and Industrial Crops of Acireale, 95024 Catania, Italy; fabiola.sciacca@crea.gov.it (F.S.); nino.virzi@crea.gov.it (N.V.); massimo.palumbo@crea.gov.it (M.P.)

[*] Correspondence: mariagrazia.melilli@cnr.it (M.G.M.); vita.distefano@unipa.it (V.D.S.); Tel.: +39-0956139916 (M.G.M.); +39-09123891948 (V.D.S.)

Received: 2 May 2020; Accepted: 5 June 2020; Published: 10 June 2020

**Abstract:** The addition of functional ingredients to breads could have effects on preventing cardiovascular diseases, cancers and inflammation. The incorporation of 0–5–10–15% of three populations of dried purslane flour on the rheological, sensorial and nutritional quality of fortified durum wheat breads were evaluated. The increase in dried purslane (up to 15%) caused an increase in the resistance to the mixture and a consequent reduction in its extensibility. The "panel test" gave a largely positive evaluation in 10% of enrichment. The fatty acids in breads resulted higher with the 5% substitution. Contrary to what has been imagined, the increase in percentage of substitution to 10 and 15% did not lead to an increase in linoleic (omega-3) and α-linolenic (omega-6) acid and probably the cause is in the cooking. The total phenols content and the antioxidant potential, evaluated by ferric reducing antioxidant potential (FRAP) and 2,2′-azinobis (3-ethylbenzothiazoline-6-sulfonic acid) (ABTS) assays of the enriched breads increased with the percentage of the dry purslane substitution. The enrichment of the durum wheat flour with 5% purslane resulted in a good compromise to obtain good rheological characteristics of loaves and breads with decreased omega-6/omega-3 ratio and good antioxidant properties.

**Keywords:** durum wheat bread; *Portulaca oleracea* L.; essential fatty acids; omega-6/omega-3 ratio; antioxidants

---

## 1. Introduction

Functional foods can be considered those whole, enriched or enhanced foods, that provide health benefits in addition to basic nutritional functions, when consumed as part of a varied diet on a regular basis.

Recent consumer interest in nutrition and health has increased the commercial demand for functional foods [1,2]. Functional foods can also be obtained by fortifying the product with so called phytochemicals; components with high nutritional value found in nature in different types of plants. Foods fortified with phytochemicals have been associated with the prevention of at least four frequent causes of death: cancer, cardiovascular diseases, high blood pressure, and diabetes [1–4]. Interests in incorporating bioactive ingredients, plant materials, herbs or spices, rich in bio-compounds, into popular foods such as bread, have grown rapidly due to the increased consumer health awareness [5]. Bread is the staple food of the Mediterranean diet and is appreciated and eaten in countries around the world. The bread produced basically from wheat flour, is rich in carbohydrate and provides more than 50% of the total energy intake. Due its relatively low cost, availability, acceptability, and widespread consumption, bread is an excellent product in which the incorporation of functional ingredients, especially omega-3 fatty acids, is attempted [6]. In recent decades, different research teams have worked on fortifying bread with natural compounds due to the demands for healthier food [7,8]. Thus, whole grains and seeds are commonly used in the production of bread [9,10]. Moreover, a technical challenge for food technologists is the production of breads with greater volume, smoother texture and good shelf life, possessing the quality characteristics derived from the functional properties of fats.

*Portulaca oleracea* L. (purslane) is an annual herbaceous plant with reddish stems and alternate leaves from the family Portulacaceae. Purslane is distributed in many parts of the world and specifically the tropical and subtropical areas. The aerial parts of the plant are somewhat crunchy, have a slight lemony taste and are consumed as salads [11,12]. It is a well known plant in traditional medicine; its medicinal value is evident from its use as purgative, cardiac tonic, emollient, muscle relaxant, anti-inflammatory and diuretic with immune-protective properties [13–15].

Purslane has been described as a power food due its high nutritive and antioxidant properties and activities, mainly acting as a free radical scavenger, metal quencher and lipid peroxidation inhibitor, thanks to its phenolic constituents and several fatty acids. Purslane is abundant in $\omega$-3 fatty acids, particularly in $\alpha$-linolenic acid (0.83 mg g$^{-1}$), for which it is considered one of the richest plant sources. Apart from $\alpha$-linolenic acid, which represents nearly up to 30% of purslane oil, other essential fatty acids have also been detected in plant tissues, such as palmitoleic, palmitic, linoleic, oleic and stearic, acids, as well as trace amounts of 20:5$\omega$-3 and 22:6$\omega$-3, namely eicosapentaenoic acid and docosahexaenoic acid, respectively [12,16]. In a number of regions in southern Italy, bread is mainly made from durum wheat, and different varieties have been identified as suitable for both bread and pasta production [17,18]. In this study, the effect of incorporating different amounts of dried purslane flour on the technological, sensorial and nutritional quality of fortified durum wheat breads was evaluated.

## 2. Materials and Methods

### 2.1. Plant Material

Purslane germplasm was collected in three different sites of eastern Sicily: Caltagirone (Cal; 37°11′07″ N-14°13′19″ W), Cassibile (Cas; 36°58′33″ N-15°12′18″ W) and Santa Venerina (SVen; 37°40′23″ N-15°19′26″ W) during June 2017. In the identified sites of collection, the plant resulted widespread and naturally covered the degraded soils. The biomorfological and chemical characterization of the plant material were reported in a previous study [12]. Linolenicacid resulted as the most abundant (0.86 mg g$^{-1}$), followed by the palmitic (0.76 mg g$^{-1}$) and oleic acids (0.25 mg g$^{-1}$) in dried whole plants [7]. The collected parts of the plant were blended thoroughly for homogeneity and washed with deionized water. After draining the excess water, the biomass (leaves + stalks) was dried at 40 °C for 3 days and ground into powder (about 300 micron). The final moisture content was less than 8%.

### 2.2. Bread-Making Test

Each form of bread was obtained adding to 400 g of commercial durum wheat flour (xg semolina + xg dried purslane) 8 g of sucrose, 8 g of salt, 24 g dehydrated mother yeast and x mL of distilled water, calculated according to the water absorption index (WA) by Brabender farinograph analysis. The obtained mixture was divided into two 200 g shapes and placed in rectangular aluminum pans of 23 × 11 × 5 cm. For each purslane population, different concentrations (5%, 10% and 15% w/w) of the substitution on the total weight of the flour were studied. Bread without purslane was used as the control (CTRL). The different types of produced bread are reported in Table 1. Three replicates were prepared for each bread sample. The bread loafs were square-shaped in an experimental aluminum box, in leavening conditions of 30 °C for 1 h, 75% r.h. and bakedin a Polin mod. Wind Pierre experimental oven (Verona. Italy), at 180 °C, for 18 min. The bread samples (Cal5, Cal10, Cal15, Cas5, Cas10, Cas 15, SVen5, SVen10 and SVen15, Table 1) were then subjected to the instrumental measurement for volume (Geass Volumometer, cm$^3$) and the height of the loaf of breads (Vernier Caliper, cm).

**Table 1.** Bread samples.

| Origin | Code | Percentage of Substitution | | |
|---|---|---|---|---|
| | | 5% | 10% | 15% |
| Cassibile | Cas | Cas5 | Cas10 | Cas15 |
| Caltagirone | Cal | Cal5 | Cal10 | Cal15 |
| S. Venerina | SVen | SVen5 | SVen10 | SVen15 |

*2.3. Rheological Characteristics*

The dough-mixing properties of the control and the different mix were examined with the Brabender Farinograph (Brabender, Duisburg, West Germany) according to the constant flour weight procedure (AACC n° 54–21). Three hundred grams' flour was mixed at the optimum water absorption and the farinograph curve was centered on the 500 BU line. According to the standard procedure, the following farinograph indices were determined: (1) water absorption of blend (WA), (2) development time of dough (DT), (3) stability of dough (S), and (4) the degree of softening of dough (DS) (Table 2). The alveographic test was used to analyze the effect of the additions on the dough rheological behavior performed by Chopin alveograph (Chopin, Villeneuve La Garenne, France) according to the standard alveographic (UNI n° 10453 method) (American Association of Cereal Chemists 2000). Each sample was analyzed in triplicate and the deformation energy W (strength) and P/L (tenacity/extensibility ratio) were calculated.

**Table 2.** Alveographic and farinographic indices of the doughs enriched with purslane averaged for concentrations and populations. Different letters indicate differences at $p < 0.05$ (according to the Duncan test).

| Populations | Alveograph Analysis | | Farinograph Analysis | | |
|---|---|---|---|---|---|
| | W (strength) ($10^{-4}$ J) | P/L (tenacity/extensibility ratio) | Water Absorption % | Development Time (min) | Stability (min) |
| | Average of concentrations | | | | |
| Cas | 86 a | 7.79 a | 64.92 c | 3.47 a | 9.35 a |
| Cal | 86 a | 7.91 a | 65.50 a | 3.45 a | 9.20 a |
| SVen | 88 a | 8.50 a | 65.05 ab | 5.45 a | 8.97 a |
| Means | 86.78 | 8.07 | 65.16 | 4.13 | 9.18 |
| | Average of populations | | | | |
| 0 | 135 a | 3.58 b | 62.90 d | 2.00 b | 4.20 d |
| 5 | 67 b | 10.15 a | 64.40 c | 2.57 b | 6.13 c |
| 10 | 58 b | 10.48 a | 65.70 b | 4.43 b | 9.47 b |
| 15 | - | - | 67.63 a | 7.50 a | 16.90 a |
| Means | 86.78 | 8.07 | 65.16 | 4.13 | 9.18 |

*2.4. Sensorial Analysis of Bread*

Bread samples were submitted to a panel of 10-trained tasters (five men and five women, aged between 27 and 60 years) in order to evaluate the sensory attributes. The panel group was an on-going panel with prior training. The panelists were selected based on their sensory skills (ability to accurately determine and communicate the sensory attributes as the appearance, odor, flavor and the texture of a product). The samples (CTRL and 5%, 10% and 15% of the fortified bread of all three populations) were served in dishes randomly labeled with three-digit random numbers for all panelists. The beads were sliced (1 cm thick) and were offered in distinct dishes at the same time. Water was provided for rinsing purposes. They were evaluated on crust thickness, elasticity, hardness, friability, apparent softness (force required for compressing the bread slice on a flat surface with a finger, to obtain a deformation about 50% of crust); on crumb elasticity, friability, cohesiveness, humidity, optical evaluation of the average size and homogeneity of the alveoli, cohesiveness to the crust. We asked the panelists to give a bread overall judgment for the overall taste and odor. To this end, a 10-point scale was used: 1—low sensation, 10—high sensation while for overall taste and odor and final overall judgment, 1corresponds to extremely unpleasant and 10 to extremely pleasant. The threshold of acceptability was set at 6.

### 2.5. Bread Colour Evaluation

Bread color data were collected with the use of a Minolta colorimeter CR, 400, as previously described by Melilli et al. [19]. The colorimeter was calibrated using the manufacturer's standard white plate (L* = 96.55; a* = −0.35; b* = −0.16), where the L* value represents light–dark spectrum with a range from 0 (black) to 100 (white), the a* value represents the green–red spectrum with a range from −60 (green) +60 (red). The b* value represents the blue–yellow spectrum with a range from −60 (blue) +60 (yellow) (CIELAB color space) [20].

### 2.6. Fatty Acids Content

The fatty acids content was determined in the germplasm collected [12], in the CTRL and in the breads enriched with purslane. The fatty acid composition was studied on 0.5 g amounts of samples after the saponification of triglycerides, as described in Melilli et al. [12], after the fatty acids were transformed into corresponding methyl esters (FAMEs) and injected in a gas chromatography–mass spectrometry (GC/MS) (Thermo-Fisher) [21,22]. Using Thermo Scientific Xcalibur Data system software for Windows, the peak areas were determined and identified by the comparison of the retention times with those of a FAMEs standard mix (Supelco CRM18918 SUPELCO FAME Mix C8–C24) separated under the same chromatographic conditions. Triplicate analyses were prepared for each sample, and the analyzed FAMEs were expressed in mg g$^{-1}$.

### 2.7. Total Phenols Content

The total phenol content in the bread samples (TPC) was determined using the Folin–Ciocalteu method as reported by Singleton, et al. [23], with some modifications [24]. One gram of fine bread powder was extracted with 10 mL of a solution MeOH:H$_2$O (80:20). After sonication for 40 min, the extracts were filtered and stored in a −20°C freezer overnight. For the determination of the TPC, 625 μL of the Folin–Ciocalteu reagent (Merck, Darmstadt, Germany, diluted 5 times) and 1.2 mL of Na$_2$CO$_3$ (7% w/v) solution were added to 125 μL of the samples' extracts. The mixtures were vortexed for two minutes and incubated in the dark for 1 h. Absorbance at 760 nm was measured using a spectrophotometer (Eppendorf Bio Spectrometer® basic). The TPC was expressed as mg gallic acid equivalent in 100 g of dried bread samples (mgGAE 100 g$^{-1}$d.m.). For every sample, the protocol was repeated three times.

### 2.8. Antioxidant Activity

Bread samples with added purslane at 5 and 10% as well as the CTRL were submitted to analysis. Two aliquots of 5 g of each dried bread sample was extracted three times with 15 mL of 70% ethanol (v/v) distilled water. After a cleanup step via centrifugation (10 min at 10,000× $g$, 4 °C), and filtration through a Millex HV 0.45 μm filter (Millipore, Billerica, MA, USA), the supernatants of each extraction cycle were recovered, combined, and used for the analysis of antioxidant activity. The antioxidant properties of the extracts from bread samples were evaluated using two antioxidant assays: the ABTS (2,2'-Azino-bis(3-ethylbenzothiazoline-6-sulfonic acid) assay and the ferric reducing antioxidant power (FRAP) assay. The ABTS assay was performed according to Re et al. [25] and the FRAP assay was performed according to Benzie and Strain [26]. The calibration curve was constructed using Trolox, an hydrophilic analogue of vitamin E. The samples were analyzed at five different dilutions, within the linearity range of the assay, as described by Gentile et al. [27]. All the measurements were repeated two times and values expressed as µmol Trolox equivalent (TE) 100 $g^{-1}$.

### 2.9. Data Analysis

Data were submitted to the Bartlett's test for the homogeneity of variance and then analyzed using an analysis of variance (ANOVA). Means were statistically separated on the basis of a Student–Newmann–Kewls test, when the 'F' test of the ANOVA for treatment was significant at least at the 0.05 probability.

## 3. Results and Discussion

### 3.1. Rheological Characteristics

The results of the rheological characteristics (Table 2) show that the origin of the purslane did not result in significant differences in the properties of the dough; on the contrary, the different percentages of substitution of wheat semolina with dried purslane induced highly significant differences both on the parameters detected by alveograph and farinograph.

It is known that gluten is composed of glutenin and gliadin subunits: the first ones give the dough toughness, while the gliadins determine its extensibility. The alveographic indices are normally related to the quality and quantity of gluten; in particular, the $p$ value indicates the dough tenacity and the L value, its extensibility. The P/L configuration ratio indicates the balance between the two factors. As shown in previous works, the P/L parameter greatly affects the bread-making quality and a value of this ratio close to unity is favorable to the baking process [17].

Table 2 shows that a partial substitution (from 5 to 10%) of wheat semolina with dried purslane induced significant modifications on the alveographic parameters (P/L and W). By comparing the CTRL sample and the formulation containing 5% dried purslane, the W value decreased by about 50%; on the contrary, the P/L value tripled. In the formulation containing 10% dried purslane, the W value decreased by about 43% and the P/L value increased, indicating inextensible dough. Moreover, when the substitution level increased, up to 15%, the alveographic indices could not be determined, indicating that the high substitution level was opposed to the dough development. The farinographic parameters included the water absorption, development time and the dough stability. Table 2 shows that the value of all the farinographic indices increased with the increase in the percentages of substitution. In particular, it shows the greatest variation of water absorption among farinograph indices, which is in agreement with the observations of Wang et al. [28]. The partial replacement, from 5 to 15% wheat semolina by the dried purslane, led to an increase in water absorption (Table 2). As for the development time, only the mixture with 15% dried purslane showed a significant increase in DT value (7.50 min). Concerning S during mixing, the addition of different percentages of dried purslane caused a progressive increase in the values, in particular in the formulation containing 15% dried purslane. Significantly, S increased from 4.2 min (no addition) to 16.90 min (15%).

### 3.2. Color, Form and Organoleptic Characteristics of the Fortified Breads

The color of the food is the first parameter of quality evaluated by consumers. The objective values of CIELAB, on crust and crumb, the height and the volume of the bread samples, subjected to sensory analyses, are reported in Table 3, as the average of the percentage of substitution and of the purslane origin. The color indices were not affected by the type of purslane added both for the crust and crumb, that on average, had L* values of 41.35 (crust) and 47.87 (crumb).

**Table 3.** L*, a*, b* in the breads enriched with purslane on the average of concentrations and populations. Different letters indicate differences at $p < 0.05$.

| Populations | Crust L* | Crust a* | Crust b* | Crumb L* | Crumb a* | Crumb b* | Height cm | Volume cm³ |
|---|---|---|---|---|---|---|---|---|
| | | | Average of concentrations | | | | | |
| Cal | 42.01 a | 9.96 a | 19.59 a | 49.00 a | 1.47 b | 16.34 a | 5.53 a | 265 a |
| Cas | 41.18 a | 9.97 a | 18.69 a | 47.57 b | 2.10 a | 15.83 a | 5.59 a | 260 a |
| SVen | 40.88 a | 10.38 a | 19.55 a | 47.04 b | 2.04 a | 16.19 a | 5.71 a | 263 a |
| Means | 41.35 | 10.10 | 19.28 | 47.87 | 1.87 | 16.12 | 5.61 | 263 |
| | | | Average of populations | | | | | |
| 0 | 46.82 a | 16.69 a | 28.09 a | 69.26 a | −0.28 d | 22.58 a | 5.67 b | 336 a |
| 5 | 39.74 c | 9.98 b | 19.02 b | 44.80 b | 2.04 c | 15.35 b | 5.98 a | 259 b |
| 10 | 37.06 d | 8.21 c | 14.55 d | 40.30 c | 2.71 b | 14.03 c | 5.48 b | 240 bc |
| 15 | 41.79 b | 5.53 d | 14.88 c | 37.11 d | 3.01 a | 12.52 d | 5.32 b | 217 c |
| Means | 41.35 | 10.10 | 19.14 | 47.87 | 1.87 | 16.12 | 5.61 | 263 |

The same behavior was noticed for the visual appearance of the loaf of breads (high 5.61 cm with a volume of 263.19 cm³ as average). Statistical differences in color were recorded in relation to the percentage of purslane substitution for the crust and crumb; in fact, together they resulted darker, increasing from 0 (CTRL) to 15%. The height of the loaf of bread resulted improved by adding 5% purslane flours.

The addition of purslane to bread was expected to influence its structure, altering the organoleptic characteristics. For this reason, the sensory properties of the bread samples were investigated in this work were addressed (Table 4). All the samples had values ranging from 8 to 6 in the overall bread judgment until 10% substitution (Table 4), except for the Cas10 sample (5.3).

**Table 4.** Sensory attributes of the crust and crumb of the fortified breads with purslane and their overall judgment. The threshold of acceptability for the bread's overall judgment is 6.

| Sensory Attributes | Crust | | | | | | | | | | Means | | | | | |
| | CTRL | Cas 5 | Cas 10 | Cas 15 | Cal 5 | Cal 10 | Cal 15 | SVen 5 | SVen 10 | SVen 15 | Cas | Cal | SVen | 5% | 10% | 15% |
|---|---|---|---|---|---|---|---|---|---|---|---|---|---|---|---|---|
| Thickness [a] | 3.0 | 4.0 | 3.0 | 4.0 | 4.0 | 4.0 | 5.0 | 4.3 | 5.0 | 4.0 | 4.0 | 4.3 | 4.4 | 4.1 | 4.0 | 4.3 |
| Elasticity [a] | 5.5 | 5.5 | 6.0 | 6.0 | 6.0 | 8.0 | 6.0 | 4.0 | 4.0 | 4.0 | 5.5 | 6.7 | 4.0 | 5.2 | 6.0 | 5.3 |
| Hardness [a] | 5.0 | 6.0 | 6.0 | 4.0 | 6.0 | 6.0 | 4.0 | 5.3 | 4.0 | 6.0 | 6.0 | 5.3 | 5.1 | 5.8 | 5.3 | 4.7 |
| Friability [a] | 4.0 | 3.0 | 2.0 | 4.0 | 4.0 | 4.0 | 4.0 | 2.0 | 2.0 | 6.0 | 3.0 | 4.0 | 3.3 | 3.0 | 2.7 | 4.7 |
| Apparent softness [a] | 3.0 | 3.0 | 2.0 | 4.0 | 4.0 | 4.0 | 3.0 | 4.0 | 2.0 | 4.0 | 3.0 | 3.7 | 3.3 | 3.7 | 2.7 | 3.7 |
| | Crumb | | | | | | | | | | | | | | | |
| Elasticity [a] | 5.0 | 6.0 | 6.0 | 4.0 | 4.0 | 6.0 | 4.0 | 3.0 | 4.0 | 6.0 | 6.0 | 4.7 | 4.3 | 4.3 | 5.3 | 4.7 |
| Apparent softness [a] | 3.3 | 4.0 | 4.0 | 4.0 | 4.0 | 4.0 | 4.0 | 4.0 | 4.0 | 4.0 | 4.0 | 4.0 | 4.0 | 4.0 | 4.0 | 4.0 |
| Friability [a] | 5.0 | 5.0 | 6.0 | 4.0 | 4.0 | 6.0 | 6.0 | 3.0 | 5.0 | 6.0 | 5.0 | 5.3 | 4.7 | 4.0 | 5.7 | 5.3 |
| Cohesiveness [a] | 5.0 | 6.0 | 6.0 | 4.0 | 4.0 | 4.0 | 4.0 | 4.0 | 4.0 | 4.0 | 6.0 | 4.0 | 4.0 | 4.7 | 4.7 | 4.0 |
| Humidity [a] | 3.8 | 5.0 | 4.0 | 4.0 | 4.0 | 4.0 | 4.0 | 3.0 | 4.0 | 6.0 | 5.0 | 4.0 | 4.3 | 4.0 | 4.0 | 4.7 |
| Average size of the alveoli [a] | 3.5 | 4.5 | 2.0 | 3.0 | 3.3 | 2.0 | 4.0 | 3.5 | 2.0 | 3.0 | 4.5 | 3.1 | 2.8 | 3.8 | 2.0 | 3.3 |
| Homogeneity of the alveoli [a] | 7.0 | 7.0 | 7.0 | 7.0 | 7.0 | 7.0 | 7.0 | 7.0 | 7.0 | 7.0 | 7.0 | 7.0 | 7.0 | 7.0 | 7.0 | 7.0 |
| Cohesiveness to the crust [a] | 7.5 | 7.0 | 6.0 | 6.0 | 6.0 | 6.0 | 6.0 | 6.0 | 6.0 | 6.0 | 7.0 | 6.0 | 6.0 | 6.3 | 6.0 | 6.0 |
| Bread overall judgment [b] | 8.0 | 8.0 | 5.3 | 4.0 | 7.3 | 6.0 | 4.0 | 8.0 | 6.7 | 4.0 | 5.8 | 5.8 | 6.2 | 7.8 | 6.0 | 4.0 |

[a] 1—good feeling and 10—bad feeling; [b] 1—extremely unpleasant, 10—extremely pleasant. CTRL—bread without purslane.

For crumb, the addition of purslane led to a darker color than for the CTRL. The thickness resulted higher than 4.0 in all the samples than in the CTRL (3.0), while the hardness sensation was higher until 10% substitution only for the Cas and Cal bread samples. SVen breads, on average of purslane fortification, gave the similar results (5.1) with respect to the CTRL (5.0).

The elasticity of crumb resulted improved in the Cas5 and Cas10 breads, while the other samples gave values lower than the CTRL (5.0). The alveoli average size and distribution were not affected by the dry purslane addition.

In general, the addition of purslane improved the darkness, as can be expected, with an increase in the percentage of substitution. With the exception of the 15% substitution, all the bread samples had good elasticity, thickness, friability and apparent softness with very small variations between the purslane accessions. The general judgment was always close to 6 (acceptability threshold) in the bread samples (means Cas 5.8, Cal 5.8 and SVen 6.2).

### 3.3. Chemical Characterization and Antioxidant Activity

Table 5 shows the FAMEs content detected in the CTRL bread and the fortified breads. The literature data of the FAMEs of dry purslane were also reported [12]. The content of the fatty acids in breads was influenced by both the percentage of substitution and the type of purslane used to enrich the bread samples. As it is possible to observe, the content in linolenic acid, linoleic acid and oleic acid is strongly influenced by the percentage of substitution. In fact, the initial content of FAMEs detected in dry purslane influenced the final fatty acid content of the fortified breads. The CTRL bread sample showed good levels of total FA ($\sum$ fatty acids 221.2 mg 100 g$^{-1}$). The results obtained showed a 54% increase in Cas5, 23% in SVen5 and 10% in Cas10. On the contrary, no increase in content was observed in all the other bread samples.

**Table 5.** Fatty acid methyl esters (FAME) content (mg 100 g$^{-1}$) in the fortified breads. Different letters indicate statistical differences at $p < 0.05$ among percentage of substitution into the same purslane population.

| | Fortified Bread Sample | | | | | | | | | | Dry Purslane * | | |
|---|---|---|---|---|---|---|---|---|---|---|---|---|---|
| | CTRL | Cal5 | Cal10 | Cal15 | Cas5 | Cas10 | Cas15 | SVen5 | SVen10 | SVen15 | Cal | Cas | SVen |
| Palmitic a. | 42.0 | 41.0 a | 35.0 b | 32.4 c | 65.1 a | 47.4 b | 29.3 c | 48.6 a | 38.5 b | 35.1 b | 39.0 | 42.0 | 35.0 |
| Oleic a. | 38.5 | 35.5 a | 30.6 b | 26.2 c | 56.9 a | 40.0 b | 24.1 c | 41.4 a | 32.9 b | 28.2 b | 20.0 | 21.0 | 10.0 |
| LA | 132.3 | 119.3 a | 104.3 b | 90.3 c | 203.0 a | 141.1 b | 83.6 c | 143.8 a | 115.1 b | 93.7 c | 88.0 | 78.0 | 51.0 |
| ALA | 8.4 | 13.2 a | 9.3b | 10.5 b | 16.6 a | 13.0 a | 7.7 b | 13.1 a | 9.5 b | 9.9 b | 84.0 | 89.0 | 67.0 |
| $\sum$ fatty acids | 221.2 | 209.0 a | 179.2 b | 159.5 c | 341.5 a | 241.5 b | 144.6 c | 246.8 a | 196.0 b | 166.9 c | 172 | 167 | 118 |
| PUFA | 140.7 | 132.5 a | 113.6 ab | 100 b | 219.6 a | 154.1 b | 91.3 c | 156.9 a | 124.6 b | 103.6 c | 231 | 230 | 163 |
| PUFA/SFA | 3.3 | 3.2 a | 3.2 a | 3.1 a | 3.4 a | 3.2 a | 3.1 a | 3.2 a | 3.2 a | 2.9 b | 4.41 | 3.98 | 3.37 |
| ω-6/ω-3 ratio | 15.7 | 9.0 b | 11.2 a | 8.6 b | 12.2 a | 10.9 b | 10.9 b | 10.9 b | 12.1 a | 9.4 b | 1.05 | 0.88 | 0.76 |

* [12]; (LA: linoleic acid; ALA: α-linolenic acid; PUFA: polyunsaturated fatty acid; SFA: saturated fatty acid).

Regarding α-linolenic acid (ALA), 8.4 mg 100 g$^{-1}$ was the concentration of this fatty acid in the CTRL, but its content was higher in the all samples analyzed, in particular in the Cal5 (13.2 mg 100 g$^{-1}$) Cas5 (16.6 mg 100 g$^{-1}$) and SVen5 (13.2 mg 100 g$^{-1}$). The linoleic acid concentration in CTRL was 132.3 mg 100 g$^{-1}$, and increased considerably in Cas5 (203.0 mg 100 g$^{-1}$) and SVen5 (143.8 mg 100 g$^{-1}$). In general, for both Cas5 and SVen5, there was an increased concentration of PUFA compared to the CTRL (140.7 mg 100 g$^{-1}$) of 219.6 and 156.9 mg 100 g$^{-1}$, respectively. Focusing the contents of ALA and LA in dry purslane [7], we would have expected an increase with higher purslane substitution, which instead was not observed. We hypothesized that the baking process may have altered the structure of the PUFAs. Regarding the ω6/ω3 ratio, which in the CTRL was 15.7 mg g$^{-1}$, this decrease was observed in all the samples and was within a range of 8.6 and 11.2 mg g$^{-1}$.

These data agree with Giaretta et al. [29], that found the addition of kinako flour and chia seed to bread resulted in a significant increase in the content of polyunsaturated fatty acid (PUFA) with a lower ω6/ω3 ratio. The effect of the baking process was mainly directed on the ω6/ω3 ratio.

The importance of the ratio of the ω6/ω3 essential FA is well examined by Simopoulos [30]. According to the author, western diets are deficient in ω3 fatty acids, and have excessive amounts of ω-6 fatty acids, promoting the pathogenesis of many diseases, including cardiovascular disease, cancer, and inflammatory and autoimmune diseases, whereas increased levels of omega-3 PUFA (a low ω-6/ω-3 ratio) exert suppressive effects.

On the contrary of FA, the TPC content, reported in Figure 1, increased with the percentage of dry purslane substitution. Common bread presented a TPC content of 27 mgGAE 100 g$^{-1}$. On the average of the three types of purslane, adding 5% dry purslane the TPC content was double (56 mgGAE100 g$^{-1}$). At 10 and 15% substitution, the effect of the dry purslane added resulted statistically different. The increase in TPC, using Cal purslane resulted 75 (Cal 10) and 91 mgGAE100 g$^{-1}$ (Cal 15), while using Cas or SVen purslane, the increase in TPC was smaller (73 mgGAE100 g$^{-1}$, value averaged at 15% of Cas and SVen substitution).

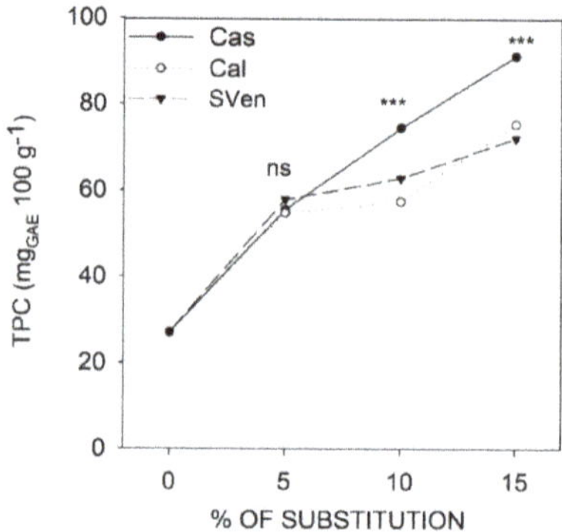


**Figure 1.** The total phenol content (TPC) of the fortified breads, obtained with durum wheat flour and purslane at three different concentrations. (ns) indicates not statistically different. (***) significant at 0.001 probability level. LSD (Least Significant Difference) ($p < 0.05$) Cas = 15.3; LSD ($p < 0.05$) Cal = 12.8; LSD ($p < 0.05$) SVen = 14.2.

Considering the results of the sensory properties of the fortified breads, the antioxidant properties, in terms of FRAP and ABTS, were analyzed until 10% purslane substitution. The FRAP (ferric reducing antioxidant power) assay was based on the reduction of ferric ion by phenols and the ABTS assay with 2,2′-azinobis(3-ethylbenzothiazoline-6-sulfonic acid) was based on free radical scavenging [16–18]. Results are graphed in Figure 2. An increase in antioxidant activity is one of the main aims of food supplementation. Both assays (FRAP and ABTS assays) highlighted the capacity of purslane to increase the antioxidant potential of enriched samples vs. the CTRL. We decided to focus the interest on the samples with the lower substitution, also in view of the best values of ALA content, TPC and antioxidant potential activity.

Regarding the ABTS results: breads with 5% purslane had an antioxidant activity of 244.7 µmol TE 100 g$^{-1}$ vs. 159.9 µmol TE 100 g$^{-1}$ in CTRL. The results of the FRAP assay had the same trend: also in this case, values did not result statistically different among the types of purslane, that on average had an antioxidant potential of 8921 vs. 580 µmol TE 100 g$^{-1}$ in CTRL.

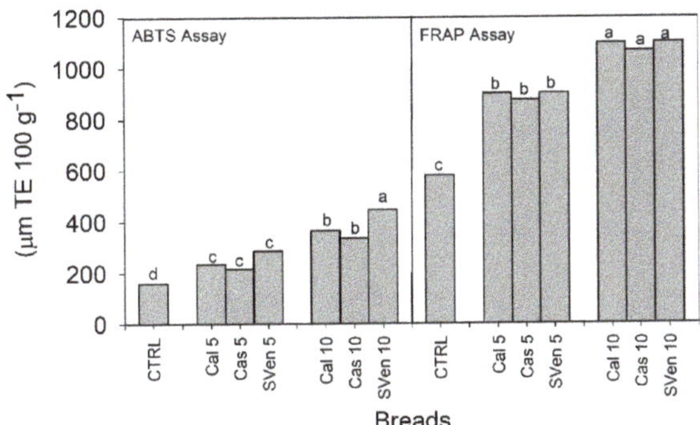



**Figure 2.** Trolox equivalent (TE) antioxidant activity (FRAP assay and ABTS assay) of the fortified breads obtained with durum wheat flour and the three populations of purslane at 5 and 10% substitution. Different letters among bars indicate significant differences at 0.05 probability level. FRAP: ferric reducing antioxidant power); ABTS 2,2′-azinobis (3-ethylbenzothiazoline-6-sulfonic acid).

## 4. Conclusions

The rheological characteristics of the wheat flour dough enriched with dried purslane provided useful information on the modifying effect of these additions on the behavior of the dough during its development, mixing and testing. However, the fundamental difference in the rheological effect of the additions was that the increase in the dose of dried purslane caused a reduction in its extensibility.

The inclusion in the wheat flour of plants rich in bioactive compounds certainly gave fortified bread good potential in relation to health benefits. Even though the increasing interest of consumers in these products, the food development and food design have to find a good compromise between the percentage of substitution of plant/herbs or spice and the sensorial properties. In our work, the enrichment of durum wheat flour by up to 5% purslane resulted in a good compromise to obtain the good rheological characteristics of loaves and breads with a decrease inomega-6/omega-3 essential FA ratio and good antioxidant properties. Considering in the Sicilian and Mediterranean tradition, common bread is made mainly using durum wheat, the addition of only 5% dry purslane could be a useful strategy to increase the bioactive compounds with potential health benefitsin bread. Trials concerning the shelf life of the products are ongoing.

**Author Contributions:** Conceptualization, M.G.M.; data curation, M.G.M., F.S., V.D.S., C.G. and M.P.; formal analysis, M.G.M., F.S., A.P., R.B. and V.D.S.; funding acquisition, M.G.M.; investigation, F.S., A.P., S.S., N.V. and C.G.; methodology, M.G.M., C.G. and M.P.; project administration, M.G.M.; resources, M.G.M.; software, N.V., C.G. and M.P.; supervision, M.G.M.; V.D.S. writing—original draft, M.G.M., V.D.S., C.G. and M.P.; writing—review and editing, M.G.M. and V.D.S. All authors have read and agreed to the published version of the manuscript.

**Funding:** This work has been supported by National Research Council, CNR—DISBA project NutrAge (project nr. 7022).

**Conflicts of Interest:** The authors declare no conflict of interest.

## References

1. Garbetta, A.; D'Antuono, I.; Melilli, M.G.; Sillitti, C.; Linsalata, V.; Scandurra, S.; Cardinali, A. Inulin enriched durum wheat spaghetti: Effect of polymerization degree on technological and nutritional characteristics. *J. Funct. Foods* **2020**, *71*, 104004. [CrossRef]

2. Padalino, L.; Costa, C.; Conte, A.; Melilli, M.G.; Sillitti, C.; Bognanni, R.; Raccuia, S.A.; Del Nobile, M.A. The quality of functional whole-meal durum wheat spaghetti as affected by inulin polymerization degree. *Carbohydr. Polym.* **2017**, *173*, 84–90. [CrossRef] [PubMed]

3. Gaudette, N.J.; Pickering, G.J. Modifying Bitterness in Functional Food Systems. *Crit. Rev. Food Sci. Nutr.* **2013**, *53*, 464–481. [CrossRef] [PubMed]

4. Msaddak, L.; Abdelhedi, O.; Kridene, A.; Rateb, M.; Belbahri, L.; Ammar, E.; Nasri, M.; Zouari, N. Opuntia ficus-indica cladodes as a functional ingredient: Bioactive compounds profile and their effect on antioxidant quality of bread. *Lipids Health Dis.* **2017**, *16*, 1–8. [CrossRef] [PubMed]

5. Ibrahim, U.K.; Salleh, R.M.; Maqsood-ul-Haque, S.N.S. Bread towards Functional Food: An Overview. *ETP Int. J. Food Eng.* **2016**, *1*, 39–43. [CrossRef]

6. Dziki, D.; Rózyło, R.; Gawlik-Dziki, U.; Świeca, M. Current trends in the enhancement of antioxidant activity of wheat bread by the addition of plant materials rich in phenolic compounds. *Trends Food Sci. Technol.* **2014**, *40*, 48–61. [CrossRef]

7. Melilli, M.; Sillitti, C.; Conte, A.; Padalino, L.; Del Nobile, M.; Bognanni, R.; Pagliaro, A. Quality characteristics of cereal based foods enriched with quinoa and inulin. In *Quinoa: Cultivation, Nutritional Properties and Effects on Health*; Peiretti, P.G., Gai, F., Eds.; Nova: Annandale, VA, USA, 2019; p. 328.

8. Sillitti, C.; Melilli, M.G.; Padalino, L.; Bognanni, R.; Tringali, S.; Conte, A.; Raccuia, S.A.; Del Nobile, M.A. Healthy pasta production using inulin from cardoon: First results of sensory evaluation. *ISHS Acta Hortic.* **2016**, *1147*, 407–412.

9. Coelho, M.S.; de las Mercedes Salas-Mellado, M. Effects of substituting chia (*Salvia hispanica* L.) flour or seeds for wheat flour on the quality of the bread. *LWT-Food Sci. Technol.* **2015**, *60*, 729–736. [CrossRef]

10. Verdú, S.; Barat, J.M.; Grau, R. Improving bread-making processing phases of fibre-rich formulas using chia (*Salvia hispanica*) seed flour. *LWT-Food Sci. Technol.* **2017**, *84*, 419–425. [CrossRef]

11. Zhou, Y.X.; Xin, H.L.; Rahman, K.; Wang, S.J.; Peng, C.; Zhang, H. Portulaca oleracea L.: A review of phytochemistry and pharmacological effects. *Biomed Res. Int.* **2015**, *2015*. [CrossRef] [PubMed]

12. Melilli, M.G.; Pagliaro, A.; Bognanni, R.; Scandurra, S.; Di Stefano, V. Antioxidant activity and fatty acids quantification in Sicilian purslane germplasm. *Nat. Prod. Res.* **2019**, 1–8. [CrossRef] [PubMed]

13. Catap, E.S.; Kho, M.J.L.; Jimenez, M.R.R. In vivo nonspecific immunomodulatory and antispasmodic effects of common purslane (*Portulaca oleracea* Linn.) leaf extracts in ICR mice. *J. Ethnopharmacol.* **2018**, *215*, 191–198. [CrossRef] [PubMed]

14. Xiu, F.; Li, X.; Zhang, W.; He, F.; Ying, X.; Stien, D. A new alkaloid from *Portulaca oleracea* L. and its antiacetylcholinesterase activity. *Nat. Prod. Res.* **2018**, 1–8. [CrossRef] [PubMed]

15. Xiang, L.; Xing, D.; Wang, W.; Wang, R.; Ding, Y.; Du, L. Alkaloids from *Portulaca oleracea* L. *Phytochemistry* **2005**, *66*, 2595–2601. [CrossRef] [PubMed]

16. Petropoulos, S.; Karkanis, A.; Fernandes, Â.; Barros, L.; Ferreira, I.C.F.R.; Ntatsi, G.; Petrotos, K.; Lykas, C.; Khah, E. Chemical Composition and Yield of Six Genotypes of Common Purslane (*Portulaca oleracea* L.): An Alternative Source of Omega-3 Fatty Acids. *Plant Foods Hum. Nutr.* **2015**, *70*, 420–426. [CrossRef] [PubMed]

17. Palumbo, M.; Spina, A.; Boggini, G. Bread-making quality of Italian durum wheat. *Ital. J. Food Sci.* **2002**, *14*, 2.

18. Spina, A.; Brighina, S.; Muccilli, S.; Mazzaglia, A.; Fabroni, S.; Fallico, B.; Rapisarda, P.; Arena, E. Wholegrain Durum Wheat Bread Fortified with Citrus Fibers: Evaluation of Quality Parameters During Long Storage. *Front. Nutr.* **2019**, *6*. [CrossRef] [PubMed]

19. Melilli, M.G.; Tringali, S.; Raccuia, S.A. Reduction of browning phenomena of minimally processed artichoke hearts. *Acta Hortic.* **2016**, *1147*, 223–236. [CrossRef]

20. Di Stefano, V.; Pitonzo, R.; Novara, M.E.; Bongiorno, D.; Indelicato, S.; Gentile, C.; Avellone, G.; Bognanni, R.; Scandurra, S.; Melilli, M.G. Antioxidant activity and phenolic composition in Pomegranate (*Punica granatum* L.) genotypes from south Italy by UHPLC/Orbitrap-MS approach. *J. Sci. Food Agric.* **2018**. [CrossRef] [PubMed]

21. Di Stefano, V.; Melilli, M.G. Effect of storage on quality parameters and phenolic content of Italian extra-virgin olive oils. *Nat. Prod. Res.* **2020**, *34*, 78–86. [CrossRef] [PubMed]

22. Grilo, F.; Novara, M.E.; D'Oca, M.C.; Rubino, S.; Lo Bianco, R.; Di Stefano, V. Quality evaluation of extra-virgin olive oils from Sicilian genotypes grown in a high-density system. *Int. J. Food Sci. Nutr.* **2020**, *71*, 397–409. [CrossRef] [PubMed]

23. Singleton, V.L.; Orthofer, R.; Lamuela-Raventós, R.M. Analysis of total phenols and other oxidation substrates and antioxidants by means of Folin-Ciocalteu reagent. *Methods Enzymol.* **1999**, *299*, 152–178.

24. Gentile, C.; Di Gregorio, E.; Di Stefano, V.; Mannino, G.; Perrone, A.; Avellone, G.; Sortino, G.; Inglese, P.; Farina, V. Food quality and nutraceutical value of nine cultivars of mango (*Mangifera indica* L.) fruits grown in Mediterranean subtropical environment. *Food Chem.* **2019**, *277*, 471–479. [CrossRef] [PubMed]

25. Re, R.; Pellegrini, N.; Proteggente, A.; Pannala, A.; Yang, M.; Rice-Evans, C. Antioxidant activity applying an improved ABTS radical cation decolorization assay. *Free Radic. Biol. Med.* **1999**, *26*, 1231–1237. [CrossRef]
26. Benzie, I.F.F.; Strain, J.J. The ferriy of plasma (FRAP) as measure of "antioxidant power": The FRAP assay. *Anal. Chem.* **1996**, *239*, 70–76.
27. Gentile, C.; Reig, C.; Corona, O.; Todaro, A.; Mazzaglia, A.; Perrone, A.; Gianguzzi, G.; Agusti, M.; Farina, V. Pomological Traits, Sensory Profile and Nutraceutical Properties of Nine Cultivars of Loquat (*Eriobotrya japonica* Lindl.) Fruits Grown in Mediterranean Area. *Plant Foods Hum. Nutr.* **2016**, *71*, 330–338. [CrossRef] [PubMed]
28. Wang, J.; Rosell, C.M.; Benedito de Barber, C. Effect of the addition of different fibres on wheat dough performance and bread quality. *Food Chem.* **2002**, *79*, 221–226. [CrossRef]
29. Giaretta, D.; Lima, V.A.; Carpes, S.T. Improvement of fatty acid profile in breads supplemented with Kinako flour and chia seed. *Innov. Food Sci. Emerg. Technol.* **2018**, *49*, 211–214. [CrossRef]
30. Simopoulos, A.P. Importance of the ratio of omega-6/omega-3 essential fatty acids: Evolutionary aspects. *World Rev. Nutr. Diet.* **2003**, *92*, 1–22. [PubMed]

*Review*

# Health Benefits and Molecular Mechanisms of Resveratrol: A Narrative Review

Xiao Meng [1], Jing Zhou [2], Cai-Ning Zhao [3], Ren-You Gan [4,*] and Hua-Bin Li [1]

[1]  Guangdong Provincial Key Laboratory of Food, Nutrition and Health, Department of Nutrition, School of Public Health, Sun Yat-Sen University, Guangzhou 510080, China; mengx7@mail2.sysu.edu.cn (X.M.); lihuabin@mail.sysu.edu.cn (H.-B.L.)

[2]  School of Public Health, Hainan Medical University, Haikou 571199, China; hy0208035@hainmc.edu.cn

[3]  Department of Clinical Oncology, Li Ka Shing Faculty of Medicine, The University of Hong Kong, Hong Kong 999077, China; zhaocn@hku.hk

[4]  Research Center for Plants and Human Health, Institute of Urban Agriculture, Chinese Academy of Agricultural Sciences, Chengdu 610213, China

*  Correspondence: ganrenyou@caas.cn; Tel.: +86-28-8020-3191

Received: 5 February 2020; Accepted: 12 March 2020; Published: 14 March 2020

**Abstract:** Resveratrol is a bioactive compound in many foods. Since its anticancer activity was reported in 1997, its health benefits have been intensively investigated. Resveratrol has antioxidant, anti-inflammatory, immunomodulatory, glucose and lipid regulatory, neuroprotective, and cardiovascular protective effects, therefore, can protect against diverse chronic diseases, such as cardiovascular diseases (CVDs), cancer, liver diseases, obesity, diabetes, Alzheimer's disease, and Parkinson's disease. This review summarizes the main findings of resveratrol-related health benefits in recent epidemiological surveys, experimental studies, and clinical trials, highlighting its related molecular mechanisms. Resveratrol, therefore, has been regarded as a potent candidate for the development of nutraceuticals and pharmaceuticals to prevent and treat certain chronic diseases.

**Keywords:** resveratrol; bioactivities; anticancer; anti-obesity; antidiabetes; molecular mechanisms

---

## 1. Introduction

Many foodstuffs and their bioactive components are beneficial to certain diseases including cardiovascular diseases (CVDs) and cancer [1,2]. Resveratrol is a polyphenol, which naturally occurs in numerous foods, such as blueberries and peanuts, as well as grapes and their derived products like red wine [3]. Since resveratrol was reported to possess strong anticancer properties in tumor initiation, promotion, and progression stages in the well-regarded journal *Science* in 1997 [4], its bioactivities and health benefits have been intensively investigated. Many epidemiologic studies demonstrated that resveratrol is effective in the prevention of some diseases, such as CVDs and cancer [5,6]. Additionally, numerous experimental studies have illustrated that resveratrol is beneficial to a broad range of diseases, including CVDs, diabetes, obesity, cancer, liver diseases, Alzheimer's disease, and Parkinson's disease through redox/inflammatory/immune signaling pathways as well as the interplay between lipid and glucose metabolism [7–14]. Driven by the promising results from experimental research, many clinical trials have also elicited the efficacy of resveratrol against certain diseases [15,16].

This review summarizes the main findings regarding the bioactivities and health impacts of resveratrol, based on systematically searching English literature from Web of Science Core Collection and PubMed in the last five years, using the key word "resveratrol". The literature was categorized into epidemiological surveys, experimental studies, and clinical trials. In particular, this review highlights the health benefits of resveratrol on chronic diseases including CVDs, cancer, liver diseases, obesity, diabetes, Alzheimer's disease, and Parkinson's disease, and its related molecular mechanisms. We hope that this narrative review paper can provide updated information about resveratrol and can attract more attention to its health benefits.

## 2. Observational Studies

Since resveratrol was illustrated to be one of the key factors in wine contributing towards the French paradox (high intake of saturated fat but low mortality from coronary heart disease), it has attracted overwhelming interest worldwide, and many epidemiological studies have investigated the relationship between resveratrol consumption and human health [17–20]. Specifically, dietary patterns rich in resveratrol were demonstrated to significantly reduce all-cause mortality [21,22]. Resveratrol also showed its potential to improve CVD risk factors, presenting significantly decreased fasting blood glucose, triglycerides (TGs), and heart rate [6]. In addition, a case-control study reported a significant inverse association between resveratrol from grapes (but not from wine) and breast cancer risk [17]. Furthermore, a lower risk of esophageal cancer was linked with higher resveratrol intake [5]. However, some null results or even harmful effects on health have also been reported. For instance, in a cross-sectional study in the Iranian population, resveratrol intake (top quantile, 0.054 mg/day and more) was positively associated with high blood pressure (hazard ratios, HR: 1.52; 95% confidence intervals, 95% CI: 1.02–2.27), without an significant association with waist circumference, TG, high-density lipoprotein (HDL), blood glucose, and metabolic syndrome [19]. Similar null outcomes were found in other studies [18,23]. In fact, the findings from different epidemiological studies are often inconsistent, because their validity depends on many factors like the study design, sample size, resveratrol dose, follow-up duration, as well as the participants' race, health status, eating patterns, and their food preference (Table 1). Nevertheless, the positive results support further exploration of what other health effects resveratrol might provide and how it achieves them.

**Table 1.** The results of resveratrol from observational studies.

| Population/Country | Study Name/Type | Sample Size (Valid Data) | Dose and Schedule | Main Findings: Resveratrol vs. Measurements/Risk Factors/Biomarkers | Ref. |
|---|---|---|---|---|---|
| Swiss/Switzerland | Case-control | N = 971 (all female; case, 369; control, 602) | Tertiles: T1: 0.0–73.0 µg/day T2: 73.1–160.7 µg/day T3: >160.7 µg/day (food-frequency questionnaire, FFQ, on weekly frequency, 2 years prior) | *Favorable:* Inversely associated with breast cancer risk (OR (95% CI): T2 vs. T1, 0.64 (0.44–0.93); T3 vs. T1, 0.55 (0.39–0.76)) | [17] |
| Iranian (Tehranian)/Iran | Cross-sectional study, part of the TLG study | N = 2618 (male, 1162; female, 1456) | Quartiles: Q1: 0.014 mg/day Q2: 0.015–0.027 mg/day Q3: 0.028–0.053 mg/day Q4: 0.054 mg/day (FFQ on a daily frequency, 1 year prior) | *Null:* Significantly associated with WC, TG, HDL, BG, and MS *Unfavorable:* The top quantile of intake (0.054 mg/day and more) was positively associated with high BP (HR = 1.52; 95% CI: 1.02–2.27) | [19] |
| Spanish/Spain | Cross-sectional study, part of the PREDIMED study | N = 1000 (male, 479; female, 521) | Quintiles: Q1: 0.48 mg/day Q2: 1.04 mg/day Q3: 2.04 mg/day Q4: 5.75 mg/day (FFQ, 1 year prior) | *Favorable:* Significantly decreased CVD risk factors (FBG (95% CI: −1.033 to −0.033); TG (95% CI: −1.998 to −0.029); and heart rate (95% CI: −0.467 to −0.087)). *Null:* Resveratrol intake was not significantly associated with TC, HDL, LDL and BP | [6] |
| Spanish/Spain | Cross-sectional study, part of the PREDIMED study | N = 7172 (male, 3249; female, 3923) | Quintiles: Q1: 0.48 mg/d Q2: 1.04 mg/d Q3: 2.04 mg/day Q4: 5.75 mg/day (FFQ, 1 year prior) | *Favorable:* High dose intake (5.75 mg/d) significantly reduced all-cause mortality by 52% (HR = 0.48; 95% CI: 0.25–0.91) *Null:* No significant CVD risk reduction (HR = 0.77; 95% CI: 0.35–1.72) | [21,22] |
| Swedish/Sweden | Case-control study | N = 1400 (case, 594 including (OAC, 181; OSCC, 158; JAC, 255)) (control, 806) | Control: 0.1 mg/day OAC: 0.07 mg/day OSCC: 0.11 mg/day JAC: 0.09 mg/day (FFQ, 20 years prior) | *Favorable:* In a significantly negative association with the risk of 3 subtypes of esophageal cancer (OAC (95% CI: 0.12–0.49); OSCC (95% CI: 0.15–0.65), and JAC (95% CI: 0.28–0.84)) | [5] |
| Italian/Italy | Cohort study, "Aging in the Chianti Region" | N = 529 (male, 236; female, 293) | Tertiles: T1: 0.1 mg/day T2: 0.1–1.1 mg/day T3: >1.1 mg/day (FFQ) | *Favorable:* Inversely associated with the risk of frailty syndrome during the first 3-year follow-up (T3 vs. T1: OR = 0.11; 95% CI: 0.03–0.45) *Null:* No substantial association with (i) risk of frailty syndrome in 6-, or 9-year follow-up; (ii) inflammatory biomarkers including IL-6, IL-1β, TNF-α, and CRP; (iii) CVD, cancer or all-cause mortality | [18] |
| Chinese/China | Cross-sectional study, | N = 1393 (male, 446; female, 947) | Mean: 0.15 mg/d (FFQ, 1 year prior) | *Null:* Not significantly associated with CVD risk factors including BP, BG, lipid profiles (TC, TG, HDL, and LDL), and carotid IMT | [23] |

Abbreviations used in the table: BG, blood glucose; BP, blood pressure; CI, confidence intervals; CRP, C-reactive protein; FBG, fasting blood glucose; FFQ, food-frequency questionnaire; HDL, high-density lipoprotein; HR, hazard ratios; IL, interleukin; IMT, intima–media thickness; JAC, (gastro-esophageal) junctional adenocarcinoma; LDL, low-density lipoprotein; MORGEN study, Monitoring Project on Risk Factors and Health in the Netherlands study; MS, metabolic syndrome; OAC, esophageal adenocarcinoma; OR: odds ratio; OSCC, esophageal squamous-cell carcinoma; PREDIMED study: Prevención con Dieta Meniterránea study; TC, total cholesterol; TG, triglyceride; TLGS: Tehran lipid and glucose study; TNF-α, tumor necrosis factor α; WC, waist circumference.

## 3. Experimental Studies

Given the observed benefits based on resveratrol consumption, a great deal of research has explored more health outcomes of resveratrol as well as the underlying molecular mechanisms.

### 3.1. Antioxidative Activities

Resveratrol has shown strong antioxidant properties in many studies [24,25]. Resveratrol protects against oxidative stress, one of the primary contributors to many diseases, through various redox-associated molecular pathways (Figure 1 and Table 2). For instance, resveratrol upregulated the phosphatase and tensin homolog (PTEN), which decreased Akt phosphorylation, leading to an upregulation of antioxidant enzyme mRNA levels such as catalase (CAT) and superoxide dismutase (SOD) [26]. Resveratrol could also improve the antioxidant defense system by modulating antioxidant enzymes through downregulation of extracellular signal-regulated kinase (ERK) activated by reactive oxygen species (ROS) [27]. Meanwhile, resveratrol reduced the ischemia-reperfusion injury-induced oxidative stress by inhibiting the activation of the p38 mitogen-activated protein kinase (MAPK) pathway, thus the levels of antioxidants like glutathione (GSH) increased, and free radicals were directly scavenged [28]. Furthermore, resveratrol activated adenosine monophosphate (AMP)-activated protein kinase (AMPK) to maintain the structural stability of forkhead box O1 (FoxO1), facilitate its translocation, and accomplish its transcriptional function [25]. Moreover, resveratrol was demonstrated to improve systemic oxidative and nitrosative stress by activating AMPK, then sirtuin 1 (SIRT1) and the nuclear factor erythroid-2-related factor 2 (Nrf2) associated antioxidant defense pathways, as Nrf2 acts as the master regulator of numerous genes encoding antioxidants and phase II-detoxifying enzymes and molecules [29,30]. Additionally, resveratrol exhibited antioxidant bioactivities by regulating antioxidant gene expression via the Kelch-like ECH-associated protein 1 (Keap1)/Nrf2 pathway and SIRT1 [31]. Recently, resveratrol was reported to attenuate oxidative injury owing to the induced autophagy via the AMPK-mediated inhibition of mammalian target of rapamycin (mTOR) signaling or via the activation of transcription factor EB (TFEB), which promoted the formation of autophagosomes and lysosomes as well as their fusion into an autolysosome [32,33]. Generally, resveratrol protects against oxidative stress mainly by (i) reducing ROS/reactive nitrogen species (RNS) generation; (ii) directly scavenging free radicals; (iii) improving endogenous antioxidant enzymes (e.g., SOD, CAT, and GSH); (iv) promoting antioxidant molecules and the expression of related genes involved in mitochondrial energy biogenesis, mainly through AMPK/SIRT1/Nrf2, ERK/p38 MAPK, and PTEN/Akt signaling pathways; (v) inducing autophagy via mTOR-dependent or TFEB-dependent pathway.

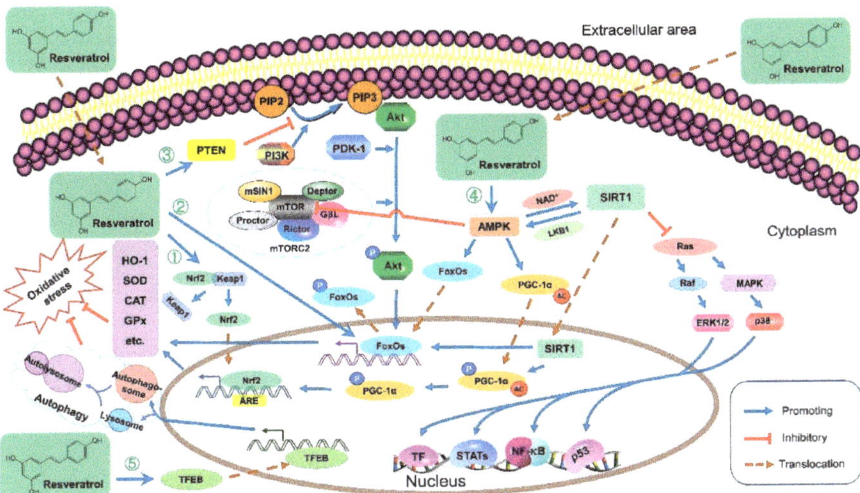


**Figure 1.** The antioxidant molecular mechanisms of resveratrol. ① Resveratrol unanchors Nrf2 in the cytoplasm, disrupting its Keap1-dependent ubiquitination and degradation. The built-up Nrf2 translocates into the nucleus, binds to ARE, and initiates the transcription of many antioxidative genes such as SOD and CAT to reduce oxidative stress. ② Resveratrol promotes the transcriptional functions of FoxOs in the nucleus to facilitate the transcription of many antioxidative genes like HO-1, contributing to the reduction of oxidative stress. ③ Resveratrol upregulated PTEN, a major antagonist of PI3K, blocking the Akt activation. Consequently, the activated Akt reduces, leading to decreased FoxOs phosphorylation. Therefore, less *p*-FoxOs translocate from the nucleus to the cytoplasm, and more FoxOs remain in the nucleus to act as transcriptional factors. ④ Resveratrol activates AMPK to maintain the structural stability of FoxOs, facilitate its translocation, and accomplish its transcriptional function. In addition, the activated AMPK phosphorylates PGC-1α, which can translocate into the nucleus, and be deacetylated by SIRT1. Then PGC-1α promotes Nrf2, leading to increased antioxidative gene expression and then reduced oxidative stress. Resveratrol activates AMPK, leading to SIRT1 activation, which inhibits MAPK signaling pathways and results in autophagy. ⑤ Resveratrol induces autophagy by activating TFEB, which promotes the formation of autophagosome and lysosome as well as their fusion into an autolysosome, leading to reduced oxidative stress. Abbreviations: AC, acetyl; Akt, protein kinase B; AMPK, AMP-activated protein kinase; ARE, antioxidant response element; CAT, catalase; ERK, extracellular signal-regulated kinase; FoxO, forkhead box protein O; GPx, glutathione peroxidase; GβL, G protein β subunit-like; HO-1, heme oxygenase 1; Keap1, Kelch-like ECH-associated protein 1; LKB1, liver kinase B1; MAP2K, mitogen-activated protein kinase kinase; MAPK, mitogen-activated protein kinase; mSIN1, mammalian stress-activated protein kinase interacting protein 1; mTOR, mammalian target of rapamycin; mTORC2, mTOR Complex 2; NAD, nicotinamide adenine dinucleotide; NF-κB, nuclear factor kappa-light-chain-enhancer of activated B cells; Nrf2, nuclear factor (erythroid-derived 2)-like 2; P, phosphorylation; p53, phosphoprotein p53; PDK1, phosphoinositide dependent kinase 1; PGC-1α, peroxisome proliferator-activated receptor gamma coactivator 1α; PI3K, phosphatidylinositol 3-kinase; PIP2, phosphatidylinositol 4,5-bisphosphate; PIP3, phosphatidylinositol-3,4,5-trisphosphate; PTEN, phosphatase and tensin homolog; Rictor, the rapamycin-insensitive companion of mTOR; SIRT1, sirtuin 1; SOD, superoxide dismutase; STAT, signal transducer and activator of transcription; TF, transcription factor; TFEB, transcription factor EB.

## 3.2. Anti-Inflammatory Activities

Resveratrol has been illustrated to have potent anti-inflammatory activities in many studies (Table 2) [34–37]. Resveratrol protected from inflammation not only by inhibiting the production of pro-inflammatory cytokines such as tumor necrosis factor α (TNF-α) and interleukin-1β (IL-1β), but also by inducing anti-inflammatory heme oxygenase-1 (HO-1) in RAW264.7 macrophages [38]. Additionally, resveratrol suppressed IL-6 transcription, modulating the inflammatory responses as an estrogen receptor $\alpha$ (ER$\alpha$) ligand mediated by SIRT1, which functions as an ER coregulator [39]. Resveratrol could also inhibit nuclear factor kappa-light-chain-enhancer of activated B cells (NF-$\kappa$B) signaling independent of SIRT1 [40]. Moreover, resveratrol attenuated inflammation by downregulating high mobility group box 1 (HMGB1) as well as suppressing NF-$\kappa$B and Janus kinase (JAK)/signal transducer and activator of transcription (STAT) signaling pathways [41,42]. In addition, both in vitro and in vivo, the anti-inflammatory effects of resveratrol were associated with its inhibition of the toll-like receptor 4 (TLR4)/NF-$\kappa$B signaling cascade [43,44]. Furthermore, resveratrol attenuated inflammation by inhibiting the activation of NACHT, LRR, and PYD domains-containing protein 3 (NALP3) inflammasome and inducing autophagy via SIRT1 upregulation and AMPK activation [34,35]. In a study on osteoarthritis, resveratrol interrupted an inflammatory amplification loop [45]. Specifically, the resveratrol-induced NF-$\kappa$B inhibition resulted in decreased IL-6 secretion, leading to suppressed signal transducer and activator of transcription 3 (STAT3) activation in macrophages. Since STAT3 was responsible for the positive regulation of IL-6 secretion, inhibition of STAT3 made IL-6 levels even lower. Resveratrol could also block ERK1/2 activation, consequently upregulating MyD88 Short, a negative regulator of inflammation [36]. Altogether, resveratrol was able to regulate the pro- and anti-inflammatory cytokines and chemokines to protect against inflammation, mainly by upregulating SIRT1, suppressing NF-$\kappa$B, and the associated cascades, as well as inhibiting NALP3 inflammasome activation.

### 3.3. Immunomodulating Effects

Resveratrol has exerted immunomodulating effects in various studies (Table 2) [46–48]. Resveratrol modulates immune response against pathogens like viruses, bacteria, and some toxic materials. For instance, resveratrol upregulated immune responses and reduced immunocyte apoptosis in chickens receiving conventional vaccinations and improved the growth of young chickens [49]. Resveratrol also reduced the activity of respiratory syncytial virus and inhibited the toll/IL-1 receptor domain-containing adaptor inducing $\beta$ interferon (TRIF) expression through upregulating sterile $\alpha$ and armadillo motif protein (SARM) [50]. In addition, resveratrol prevented enterovirus 71 (EV71) replication and reduced the virus-induced elevated IL-6 and TNF-$\alpha$ secretion in rhabdosarcoma cells via suppressing I$\kappa$B kinase (IKK)/NF-$\kappa$B signaling pathway [48]. Moreover, resveratrol inhibited human rhinoviruses-16 replication and normalized virus-induced IL-6, IL-8 and regulated on activation normal T cell expressed and secreted (RANTES) as well as the expression of intercellular adhesion molecule-1 (ICAM-1) [47]. Additionally, resveratrol maintained the immune function in rotavirus-infected piglets, resulting in attenuated diarrhea and inflammation [51]. Moreover, resveratrol triggered an immune response to protect against non-typeable *Haemophilus influenzae* (a respiratory bacterium) without developing resistance in vitro [46]. Resveratrol also decreased bacterial viability and reduced infectious airway inflammation without noticeable host toxicity in vivo [46,52]. In addition, resveratrol showed immunomodulatory properties via reducing bacterial and inflammatory biomarkers in lipopolysaccharides (LPS)-challenged primary Atlantic salmon macrophages [52]. Resveratrol also modulated immunity caused by some toxic materials like concanavalin A (Con A), showing upregulation of SIRT1 and reduction of cytokines such as TNF-$\alpha$, interferon $\gamma$ (IFN-$\gamma$), IL-6, and monocyte chemoattractant protein-1 (MCP-1) [53]. Interestingly, resveratrol was found to strongly enhance immune activity in immunosuppressive mice, showing a bidirectional regulatory effect on immunity [54]. Specifically, resveratrol improved spleen lymphocyte proliferation, enhanced the function of peritoneal macrophages, and increased the CD4$^+$ cells in peripheral blood. Furthermore, some cytokines in the serum were upregulated, such as IL-1$\alpha$/$\beta$, IL-2, and TNF-$\alpha$. Based on a statistical analysis of human microarray data, a recent study revealed that resveratrol regulated many immune response pathways including peroxisome proliferator-activated receptor $\alpha$ (PPAR-$\alpha$)/retinoid X receptor $\alpha$ (RXR$\alpha$) activation, IL-10 signaling, natural killer cell signaling, leucocyte extravasation signaling, and IL-6 signaling [55]. Recently, it was revealed that resveratrol could suppress the aryl hydrocarbon receptor (AhR) pathway, resulting in the reversal of imbalanced Th17/Treg, the main characteristic of immune thrombocytopenic purpura [56].In short, resveratrol could modulate both cellular and humoral immunity to reduce replication and the viability of pathogens, and bidirectionally regulate the related cytokine/chemokine production through the canonical immune response pathways as mentioned above.

### 3.4. Cardiovascular Diseases

Resveratrol has been reported to protect against CVDs in certain research (Table 2) [57–59]. Resveratrol prevented the pathological progression of hypertension, a major risk factor of CVDs, through Nrf2 activation, owing to its antioxidant and anti-inflammatory capacity [60]. Resveratrol could also lower blood pressure in hypertensive mice by inducing oxidative activation of cyclic guanosine monophosphate (cGMP)-dependent protein kinase 1$\alpha$ (PKG1$\alpha$) [61]. Atherosclerosis is another main contributor to CVDs. Resveratrol was able to block atherosclerotic plaque progression by acting against pro-atherogenic oxysterol signaling in M1 (inflammation-encouraging) and M2 (inflammation-decreasing) macrophages [57]. Meanwhile, resveratrol prevented the activation of inflammasome, a contributor to the vascular inflammatory injury and atherosclerosis, via downregulating NF-$\kappa$B p65 and p38 MAPK expression, and upregulating SIRT1 expression [62]. In addition, resveratrol ameliorated atherosclerosis partially through restoring intracellular GSH via AMPK-$\alpha$ activation, resulting in inhibited monocyte differentiation and reduced pro-inflammatory cytokine production [59]. Moreover, resveratrol regulated the band 4.1, ezrin, radixin, and moesin (FERM)-kinase and Nrf2 interaction, leading to decreased expression of ICAM-1 and then the inhibition of monocyte adhesion [63]. Resveratrol also exhibited antithrombotic effects via decreasing the tissue factors like TNF-$\alpha$, and such action could be facilitated by aortic endothelial cells that could deconjugate resveratrol metabolites to free resveratrol [64]. Furthermore, one of the atherosclerosis consequences, high fat/sucrose diet-induced central arterial wall stiffening, was improved by resveratrol based on its protective activities against oxidative stress and inflammation [7]. Resveratrol also effectively prevented CVD by improving the cardiac and vascular autonomic function, protecting the erythrocytes via interacting with hemoglobin and reducing heme-iron oxidation [65,66]. In a heart failure model, resveratrol mitigated atrial fibrillation by upregulating PI3K and endothelial NOS (eNOS) [8]. In summary, the cardiovascular protective effects of resveratrol mainly depend on the capabilities of reducing oxidative stress and alleviating inflammation through Nrf2 and/or SIRT1 activation, PI3K/eNOS upregulation, and NF-$\kappa$B downregulation.

### 3.5. Cancers

Resveratrol has exhibited protective impacts on various cancers, like colorectal cancer, lung cancer, breast cancer, prostate cancer, ovarian cancer, cervical cancer, liver cancer, and gastric cancer (Table 2) [14,67–71]. For instance, resveratrol was reported to inhibit the formation and growth of colorectal cancer by downregulating oncogenic KRAS expression [68]. Resveratrol also prevented tumorigenesis and progression of non-small cell lung cancer (NSCLC) cells by interrupting glycolysis via inhibition of hexokinase II expression, which was mediated by downregulation of the epidermal growth factor receptor (EGFR)/Akt/ERK1/2 signaling pathway [69]. Moreover, resveratrol showed pro-apoptotic/anti-proliferative effects in LNCaP cells (human prostate adenocarcinoma cells) through inducing the expression of cyclooxygenase (COX)-2, promoting ERK1/2 activation, and facilitating p53-dependent anti-proliferation gene expression [14]. In addition, resveratrol could decrease the efficiency of ovarian cancer cells adhering to peritoneal mesothelium in vitro by downregulating the production of $\alpha$5$\beta$1 integrins and upregulating the release of soluble hyaluronic acid [70]. Resveratrol was also reported to inhibit the expression of phospholipid scramblase 1 (PLSCR1), leading to the growth inhibition of HeLa cells [71]. Furthermore, resveratrol showed proliferation-inhibitory and apoptosis-inducing effects in HepG2 cells by activating caspase-3 and caspase-9, upregulating the

Bax/Bcl-2 ratio, and inducing p53 expression [72]. Resveratrol also inhibited the invasion and migration of human gastric cancer cells by blocking the epithelial-to-mesenchymal transition, which was mediated by metastasis-associated lung adenocarcinoma transcript 1 (MALAT1) [73]. Additionally, resveratrol protected against breast cancer metastasis by promoting antitumor immune responses via blunting STAT3, leading to inhibited generation and function of tumor-evoked regulatory B cells (tBregs) as well as decreased production of transforming growth factor $\beta$ (TGF-$\beta$) (a downstream target of STAT3), which was required by the tBregs to convert resting $CD4^+$ T cells to the metastasis-promoting $FoxP3^+$ regulatory T cells (Tregs) [67]. Although some research showed that resveratrol may be beneficial in breast cancer chemoprevention due to its non-estrogen function [74], different doses of resveratrol showed controversial effects due to its interaction with estrogens, which induce cellular proliferation and play a key role in breast cancer development and progression. Specifically, it was found that high concentrations of resveratrol could inhibit the proliferation of estrogen receptor alpha positive (ER$\alpha$+) breast cancer, while low concentrations increased the growth of ER$\alpha$+ cells [75,76]. It was reported that at low concentrations, resveratrol not only bound to ERs due to its structural similarity with E2, but also increased the formation of estrogen precursor steroids and inhibited the inactivation of active steroids, resulting in elevated active estrogen levels, leading to breast cancer cell growth and progression [77,78]. However, "more resveratrol is better" was challenged in some cases [79]. For instance, in terms of colorectal cancer chemoprevention, lower doses of resveratrol from dietary exposures exerted a better efficacy than high doses (200 times higher previously used in phase I clinical trials) due to its pro-oxidant activity at high doses and AMPK signaling upregulation.

Collectively, resveratrol has shown its anticancer bioactivities by impairing glycolysis, inhibiting cancer cell growth and proliferation, inducing apoptosis, promoting antitumor immune responses, and preventing adhesion, migration and invasion of cancer cells by modulating related molecules and gene expression through various signaling pathways. Of note, different doses may lead to very different effects, which sometimes could be opposite. Therefore, consideration regarding doses and matrix should be paid more attention in future studies.

### 3.6. Liver Diseases

Resveratrol has shown its protective impacts on several liver diseases in some studies (Table 2) [9,80–82]. Specifically, resveratrol alleviated non-alcoholic fatty liver disease (NAFLD) by upregulating the low-density lipoprotein receptor (LDLR) and scavenger receptor class B type I (SRB1) gene expressions in the liver [83], or by regulating autophagy and decreasing the activity of NF-$\kappa$B, resulting from restoring its inhibitor, nuclear factor of kappa light polypeptide gene enhancer in B-cells inhibitor $\alpha$ (I$\kappa$B$\alpha$) [84]. Resveratrol also improved high-fat diet (HFD)-induced fatty liver by downregulating adipose differentiation-related proteins and increasing the numbers of $CD68^+$ Kupffer cells [9]. As for chemical-induced liver diseases, resveratrol could markedly restore the morphology and function of alcohol-injured liver through inducing autophagy, or downregulating hypoxia-inducible factor 1$\alpha$ (HIF-1$\alpha$) expression [85,86]. In addition, resveratrol ameliorated $CCl_4$-induced liver injury by blocking the Notch signaling pathway [82]. Furthermore, resveratrol attenuated N′-Nitrosodimethylamine-induced hepatic fibrosis by restoring the architecture and normalizing collagen deposition, mainly due to its antioxidative activities and downregulation of smooth muscle actin ($\alpha$-SMA), which suppressed hepatic stellate cell activation [80,81]. Moreover, resveratrol pretreatment mitigated liver cirrhosis by improving the homing of bone marrow-derived mesenchymal stem cells [87]. In summary, resveratrol could improve NAFLD, chemical-induced liver injuries, fibrosis, and cirrhosis by modulating redox status, regulating lipid metabolism, ameliorating inflammation, and inducing autophagy with various cytokines, chemokines, and transcription factors involved.

### 3.7. Diabetes

Resveratrol has been elicited to attenuate diabetes and its relevant complications in many studies (Table 2) [88–91]. Resveratrol was observed to significantly reduce blood glucose levels, plasma lipids, and free fatty acids in diabetic mice, and it inhibited the expression of inflammatory mediators (e.g., ICAM-1, vascular cell adhesion molecule-1, and MCP-1) both in the aorta and in the blood, by inhibiting the NF-κB pathway [92]. In addition, resveratrol could relieve diabetes via increasing insulin action and glucose utilization due to visfatin expression restoration, SIRT1 activation, and glucose transporter modulation [89]. Moreover, resveratrol increased glucose uptake to improve insulin resistance in the muscle by decreasing diacylglycerol (DAG) accumulation and protein kinase C θ (PKC-θ) translocation, and preventing lipolysis under the condition of adipose hypoxia, because resveratrol could preserve phosphodiesterase 3B expression (PDE 3B) to downregulate cyclic adenosine monophosphate (cAMP), leading to the inhibition of protein kinase A (PKA)/hormone-sensitive lipase (HSL) activation [90]. Moreover, resveratrol showed protective effects on adipose tissue in diabetic mice by preventing ROS-mediated mitochondrial fission via AMPK-dependent upregulation of Drp1 phosphorylation, and by blocking the activation of NALP3 inflammasome via inhibition of endoplasmic reticulum stress (ERS) [93]. Resveratrol also protected against diabetic complications such as myocardial fibrosis, diabetic nephropathy, and erectile dysfunction [11,88,94]. Furthermore, maternal resveratrol administration to the rats was evidenced to prevent the offspring's glucose intolerance and islet dysfunction, which were associated with gestational diabetes [91]. In summary, resveratrol could effectively regulate glucose metabolism, improve insulin resistance, improve diabetic complications, and restore the function of multiple systems via modulating SIRT1/NF-κB/AMPK signaling pathways and some associated molecules like NALP3 inflammasome, as well as the expressions of relevant genes.

### 3.8. Obesity

Obesity has become a severe health issue globally. Resveratrol significantly decreased the body weight and fat mass in mice with HFD-induced obesity, showing reduced leptin and lipids in plasma, modulated metabolism of glucose and insulin, and restored immune dysfunction, via the activation of PI3K/SIRT1 and Nrf2 signaling pathways, and the inhibition of transcriptional regulators (e.g., EP300 gene), which are involved in the differentiation of adipocytes as well as lipid storage and metabolism [95,96]. Moreover, besides a significant dose-dependent decrease of weight gain and lipid deposition in the liver and adipose tissues of HFD-induced obese mice, low concentrations of resveratrol (1–10 μM) suppressed adipogenic differentiation in pre-adipocytes, downregulated the expression of peroxisome proliferator-activated receptor γ (PPAR-γ) and perilipin protein in differentiated adipocytes, and inhibited TNF-α-induced lipolysis in mature adipocytes [97]. Additionally, resveratrol prevented against obesity through markedly enhancing the catecholamine production, accompanied by suppressing the pro-inflammatory M1 macrophages and activating anti-inflammatory M2 macrophages in white adipose tissue, which play a pivotal role in the trans-differentiation of white adipocytes into beige adipocytes [98]. Furthermore, resveratrol administrated to the pregnant and lactating mice led to promoted white adipose browning and thermogenesis in the male descendants, and these health benefits persisted and prevented obesity in their future life [99]. In addition, resveratrol protected against sarcopenic obesity by improving mitochondrial function and reducing oxidative stress through the PKA/liver kinase B1 (LKB1)/AMPK pathway [100]. Resveratrol also showed positive impacts on obesity-related complications, such as reproductive dysfunction like infertility and endocrine disorders [101,102]. To summarize, resveratrol has been illustrated to decrease body weight, regulate lipid deposition, modulate adipocyte gene expression, and promote white adipose browning, via PI3K/SIRT1, Nrf2, PPAR-γ, TNF-α, and PKA/LKB1/AMPK signaling pathways (Table 2).

### 3.9. Alzheimer's Disease and Parkinson's Disease

Alzheimer's disease and Parkinson's disease are neurodegenerative disorders, seriously decreasing life quality, while resveratrol may have the potential to improve these diseases. For instance, resveratrol inhibited the aggregation of amyloid β (Aβ), a key factor in Alzheimer's disease, by modulating specific proteins such as ubiquitin-like protein (UBL)/X-box binding protein 1 (XBP-1) involved in proteostasis [103]. Furthermore, resveratrol prevented memory loss in Alzheimer's disease by decreasing elevated levels of mitochondrial complex IV protein in the mouse brain via the activation of SIRT1 and AMPK pathways [104,105]. In terms of Parkinson's disease, resveratrol ameliorated ERS by downregulating the gene expression of C/EBP homologous protein (CHOP) and glucose-regulated protein 78 (GRP78), inhibiting caspase-3 activity in the rat brain, and ameliorating oxidative damage via suppressing xanthine oxidase activity and protein carbonyl formation as well as activating the glutathione peroxidase and Nrf2 signaling pathway [10]. Resveratrol also alleviated Parkinson's disease through elevating miR-214 expression, leading to decreased mRNA expression of $\alpha$-synuclein [106]. Taken together, resveratrol ameliorated Alzheimer's disease and Parkinson's disease by activating the SIRT1, AMPK, and Nrf2 signaling pathways and modulating the associated gene expressions (Table 2).

### 3.10. Sex-Dependent Effects of Resveratrol

Acting as an estrogen agonist, resveratrol showed sex-dependent effects on some diseases, which causes increasing concerns (Table 2). CVD risk increases with increasing age, gradually in men while disproportionately in women, and such a lower risk was in association with estrogen's cardioprotective properties [107,108]. In a recent in vivo study, sex differences were observed in rats with surgically-induced myocardial infarction (MI) due to the resveratrol treatment (2.5 mg/kg/d) [109]. Superior improvements were observed in females in terms of interventricular septal wall dimension at systole (IVSDs), end-systolic volume (ESV), ejection fraction (EF), fractional shortening (FS), and isovolumic relaxation time (IVRT), among which IVRT was purely sex-dependent. In another study, long-term resveratrol treatment in rats (50 mg/L in drinking water, 21 days) increased the relaxations to estrogen in aortae, more potently in males, probably due to the effects of resveratrol on promoting nitric oxide and/or suppressing superoxides [110]. Alongside this, it was revealed that resveratrol (20 mg/kg) significantly increased dopamine transporter (DAT) in the striatum in female but not in male mice [111]. The in vitro study in the same research indicated that resveratrol upregulated DAT in the dopaminergic cells by inducing its gene transcription. Additionally, sex differences in the effect of resveratrol were also found in a mouse model of dextran sulfate sodium-induced colitis [112]. Adverse effects were observed in females but not in males, regarding weight loss, stool consistency, and discomfort. Such results indicate that special attention should be paid to the application of resveratrol, a phytoestrogen, which can interact with hormone receptors and result in sex-dependent effects that can be beneficial or harmful.

**Table 2.** Bioactivities and potential mechanisms of resveratrol from experimental studies.

| Study Type | Subject | Dose | Main Findings | Ref. |
|---|---|---|---|---|
| *Antioxidative activities* | | | | |
| In vitro | HepG2 cells | 0–100 µM | Dose-dependently increasing antioxidant effects by enhancing SIRT2's activity to deacetylate Prx1 | [24] |
| In vitro | HepG2, C2C12, and HEK293 cells | 10, 25 µM | Activating AMPK to maintain the structural stability of FoxO1 | [25] |
| In vitro | MCF-7 cells | 1 nM, 0.02 µM, 0.1 µM, 0.5 µM, 1.5 µM | Upregulating PTEN (except at the highest dose, 1.5 µM), which decreased Akt phosphorylation, leading to an upregulation of antioxidant enzyme mRNA levels such as CAT and SOD | [26] |
| In vivo | Rats | 20 mg kg/b.w./day | Improving the antioxidant defense system by modulating antioxidant enzymes through downregulation of ERK activated by ROS | [27] |
| In vivo | Rats | 10 mg/kg b.w. | Reducing the ischemia-reperfusion injury-induced oxidative stress by inhibiting the activation of p38 MAPK pathway to increase antioxidants like GSH and scavenge free radicals | [28] |
| In vivo | Rats | 5, 10 mg/kg | Activating SIRT1 to scavenge ROS | [29] |
| In vivo | Mice | 15, 30, 60 mg/kg | Activating AMPK, SIRT1, and Nrf2 associated antioxidant defense pathways to improve systemic oxidative and nitrosative stress | [30] |
| In vivo | Sows | 300 mg/kg | Regulating antioxidant gene expression via Keap1/Nrf2 pathway and SIRT1 | [31] |
| In vitro | HUVECs | 10 µM | Inducing autophagy via the activation of TFEB | [32] |
| In vitro | HEK293 cells or HEK293T | 5 µg/mL | Inducing autophagy via the AMPK-mediated inhibition of mTOR signaling | [33] |
| *Anti-inflammatory activities* | | | | |
| In vivo | Mice | 8 mg/kg/day | Inhibiting the activation of NALP3 inflammasome and inducing autophagy via SIRT1 upregulation | [34] |
| In vitro | J774 mouse macrophages, Mouse bone-marrow cells | 0.5–100 µM | Inhibiting the activation of NALP3 inflammasome | [35] |
| In vitro; In vivo | BEAS-2B cells, Mice | 25 µM, 20 mg/kg | Inducing NF-κB inhibition, decreasing IL-6 secretion, suppressing STAT3 activation, blocking ERK1/2 activation, and upregulating MyD88 Short | [36] |
| In vitro | RAW264.7 macrophages | 0–20 µM | Inhibiting the production of pro-inflammatory cytokines, such as TNF-α and IL-1β, but also by inducing anti-inflammatory HO-1 | [38] |
| In vitro | RAW264.7 macrophages, MCF-7 cells | 10 µM | Suppressing IL-6 transcription, modulating the inflammatory responses as an ERα ligand mediated by SIRT1. | [39] |
| In vitro | Mouse C2C12 myoblasts | 20, 50, 100 µM | Inhibiting NF-κB signaling independent of SIRT1 | [40] |

**Table 2.** *Cont.*

| Study Type | Subject | Dose | Main Findings | Ref. |
|---|---|---|---|---|
| In vitro | RAW264.7 macrophages | 1, 5, 10, 20, 40 μM | Downregulating HMGB1 as well as suppressing NF-κB and JAK/STAT signaling pathways | [41] |
| In vitro | U937 monocytic cells | 15, 30, 50 μM | Inhibiting NF-κB and JAK/STAT signaling pathways | [42] |
| In vitro<br>In vivo | NRK-52E,<br>Rat | 100 μmol/mL,<br>0.23 μg/kg | Inhibiting TLR4/NF-κB signaling cascade | [43] |
| In vivo | Rats | 30, 10 and 3 mg/kg, | Inhibiting TLR4/NF-κBp65/MAPKs signaling cascade | [44] |
| In vitro | Primary chondrocytes and macrophages | 10, 25, 50, 100 μM | Interrupting an inflammatory amplification loop | [45] |
| *Immunomodulating effects* | | | | |
| In vitro | A549 cells | 56.25, 112.5 μg/mL | Triggering an immune response to protect against non-typeable *Haemophilus influenzae* without developing resistance | [46] |
| In vitro | H1HeLa cells, Human nasal epithelia | 0–300 μM | Inhibiting human rhinoviruses-16 replication and normalized virus-induced IL-6, IL-8, and RANTES as well as the expression of ICAM-1 | [47] |
| In vitro | Rhabdosarcoma cells | 2.5–100 μg/mL | Preventing EV71 replication, reducing the virus-induced elevated IL-6 and TNF-α secretion via suppressing IKK/NF-κB signaling pathway | [48] |
| In vivo | Chickens | 200, 400, 800 mg/kg | Reducing immunocyte apoptosis in chickens receiving conventional vaccinations, and improving the growth of young chickens | [49] |
| In vivo | Piglets | 3, 10, 30 mg/kg/d | Maintaining the immune function and attenuating diarrhea and inflammation | [51] |
| In vitro | Atlantic salmon macrophages | 10, 30, 50 μM | Reducing bacterial and inflammatory biomarkers in LPS-challenged primary Atlantic salmon macrophages | [52] |
| In vivo | Mice | 30 mg/kg | Upregulating SIRT1 and reducing cytokines such as TNF-α, IFN-γ, IL-6, and MCP-1 | [53] |
| In vivo | Mice | 30 mg/kg | Enhancing immune activity in immunosuppressive mice, showing a bidirectional regulatory effect on immunity | [54] |
| In vitro | Human CD4+ T cells | 10, 30, or 50 μM | Suppressing the AhR pathway, resulting in the reversal of imbalanced Th17/Treg | [56] |
| *Cardiovascular diseases* | | | | |
| In vivo | Rhesus monkeys | 80 mg/day (1st year), 480 mg/day (2nd year) | Improving central arterial wall stiffening based on its antioxidative and anti-inflammation | [7] |
| In vivo | Rabbits | 2.5 mg/kg | Mitigating atrial fibrillation by upregulating PI3K/AKT/eNOS | [8] |
| In vitro | Peripheral blood mononuclear cells | 3–80 μM | Blocking atherosclerotic plaque progression by acting against pro-atherogenic oxysterol signaling in M1 and M2 macrophages | [57] |

**Table 2.** *Cont.*

| Study Type | Subject | Dose | Main Findings | Ref. |
|---|---|---|---|---|
| In vitro<br>In vivo | THP-1 monocytes,<br>Mice | 0, 25, 50, 100 µM (dose-dependent),<br>10 mg/kg/day | Ameliorating atherosclerosis partially through restoring intracellular GSH via AMPK-α activation, inhibiting monocyte differentiation, and reducing pro-inflammatory cytokine production | [59] |
| In vivo | Rats | 50 mg/L | Preventing the pathological progression of hypertension through Nrf2 activation | [60] |
| In vitro;<br>In vivo | Rat aortic smooth muscle cells;<br>Mice | 100 µmol/L,<br>~320 mg/kg | Lowering blood pressure by inducing oxidative activation of cGMP-dependent PKG1α | [61] |
| In vivo | Rats | 50 mg/kg/day | Preventing the activation of inflammasome via downregulating NF-κB p65 and p38 MAPK expression, and upregulating SIRT1 expression | [62] |
| In vivo | Mice | 20 mg/kg | Regulated the FERM-kinase and Nrf2 interaction, decreasing the expression of ICAM-1, and inhibiting monocyte adhesion | [63] |
| In vivo | Rats | 1.24 µg/d | Improving the cardiac and vascular autonomic function | [65] |
| In vitro | Human RBCs | 100 µM | Protecting the erythrocytes via interacting with hemoglobin and reducing heme-iron oxidation | [66] |
| *Cancers* | | | | |
| In vitro | LNCaP cells | 5, 10, 20, 50 µM | Inducing the expression of COX-2, promoting ERK1/2 activation, and facilitating p53-dependent anti-proliferation gene expression | [14] |
| In vitro;<br>In vivo | tBregs;<br>Mice | 12.5 µM;<br>20, 50, 500 µg/mouse | Preventing breast cancer metastasis by promoting antitumor immune responses via blunting STAT3, leading to inhibited generation and function of tBregs as well as decreased production of TGF-β | [67] |
| In vivo | Mice | 150, 300 ppm | Inhibiting the formation and growth of colorectal cancer by downregulating oncogenic KRAS expression | [68] |
| In vitro;<br>In vivo | NSCLC cells<br>Mice | 25, 50, 100 µM,<br>30 mg/kg every 3 days | Preventing tumorigenesis and progression by interrupting glycolysis via inhibition of hexokinase II expression, which was mediated by downregulation of EGFR/Akt/ERK1/2 signaling pathway | [69] |
| In vitro | MCF-7 cells<br>MVLN cells | Low: 0.1 and 1 µM; High: 10 and 25 µM; | Low concentrations: Increasing the growth of ERα+ cells<br>High concentrations: Inhibiting the proliferation of eERα+ breast cancer | [75] |
| In vitro | KPL-1, MCF-7, MKL-F cells | Low (KPL-1, ≤22 µM; MCF-7, ≤4 µM); High: ≥44 µM | Low concentrations: Causing cell proliferation ER+ cells<br>High concentrations: Suppressing cell growth | [76] |
| In vitro<br>In vivo | Apc10.1 cells;<br>Mice;<br>Humans | 0.001–1 µM;<br>0.7, 14.3 mg/kg diet;<br>5 mg, 1 g | Lower doses of resveratrol: Showing superior efficacy than high doses due to the pro-oxidant activity and AMPK signaling upregulation | [79] |

**Table 2.** *Cont.*

| Study Type | Subject | Dose | Main Findings | Ref. |
|---|---|---|---|---|
| In vitro | A2780, OVCAR-3, SKOV-3 cells | 10, 50, 100 µM | Decreasing the efficiency of ovarian cancer cells adhering to peritoneal mesothelium by downregulating the production of $\alpha5\beta1$ integrins and upregulating the release of soluble hyaluronic acid | [70] |
| In vitro | Hela cells | 0.1, 1, 10 µM, 10, 20, 50, 100 µM | Inhibiting the expression of PLSCR1, leading to the growth inhibition of HeLa cells | [71] |
| In vitro | HepG2 cells | 25, 50, 100, 200 µM | Inhibiting proliferation and inducing apoptosis by activating caspase-3 and caspase-9, upregulating the Bax/Bcl-2 ratio, and inducing p53 expression | [72] |
| In vitro | SGC7901 and BGC823 cells | 5, 10, 25, 50, 100, 200, and 400 µM | Inhibiting the invasion and migration of human gastric cancer cells by blocking the MALAT1-mediated epithelial-to-mesenchymal transition | [73] |
| *Liver diseases* | | | | |
| In vivo | Mice | 0.2% of diet | Improving HFD-induced fatty liver by downregulating adipose differentiation-related proteins and increasing the numbers of $CD68^+$ Kupffer cells | [9] |
| In vivo | Rats | 10 mg/kg | Attenuating hepatic fibrosis by restoring the architecture and normalizing collagen deposition, mainly due to its antioxidative activities and downregulation of $\alpha$-SMA | [80] |
| In vivo | Rats | 50, 100 mg/kg | Alleviating NAFLD by upregulating LDLR and SRB1 gene expressions | [83] |
| In vivo | Rats | 250 mg/kg/day | Downregulating HIF-1$\alpha$ expression and mitochondrial ROS production | [85] |
| In vitro; In vivo | HepG2 cells; Mice | 45 µmol 10, 30, 100 mg/kg | Restoring the morphology and function of alcohol-injured liver through inducing autophagy | [86] |
| In vivo | Rats | 10 mg/kg | Mitigating liver cirrhosis by improving the homing of bone marrow-derived mesenchymal stem cells | [87] |
| *Diabetes* | | | | |
| In vivo | Rats | 20 mg/kg | Increasing insulin action and glucose utilization due to visfatin expression restoration, SIRT1 activation, and glucose transporter modulation | [89] |
| In vivo | Mice | 50 mg/kg | Increasing glucose uptake to improve insulin resistance in the muscle by decreasing DAG accumulation and PKC-$\theta$ translocation, and preventing lipolysis under the condition of adipose hypoxia | [90] |
| In vivo | Rats | 147.6 mg/kg/day | Preventing the offspring's glucose intolerance and islet dysfunction | [91] |
| In vivo | Mice | 0.3% diet | Reducing blood glucose levels, plasma lipids, and free fatty acids, inhibiting the expression of inflammatory mediators both in the aorta and in the blood, by inhibiting the NF-κB pathway | [92] |

**Table 2.** *Cont.*

| Study Type | Subject | Dose | Main Findings | Ref. |
|---|---|---|---|---|
| In vivo | Mice | 50 mg/kg | Preventing ROS-mediated mitochondrial fission via AMPK-dependent upregulation of Drp1 phosphorylation, and blocking the activation of NALP3 inflammasome via inhibition of ERS | [93] |
| *Obesity* | | | | |
| In vivo | Zebrafish | 40 mg/kg/day | Inhibiting transcriptional regulators such as EP300 | [95] |
| In vivo | Mice | 0.06% diet | Decreasing the body weight and fat mass, reducing leptin and lipids in plasma, modulating metabolism of glucose and insulin, and restoring immune dysfunction by activating PI3K/SIRT1 and Nrf2 signaling pathway | [96] |
| In vitro; In vivo | 3T3-L1 cells; Mice | 0.03 to 100 µM; 1, 10, 30 mg/kg | In vitro: low concentrations of resveratrol (1-10 µM) suppressed adipogenic differentiation in pre-adipocytes, downregulated the expression of PPAR-γ and perilipin protein in differentiated adipocytes, and inhibiting TNF-α-induced lipolysis in mature adipocytes. In vivo: Dose-dependently decreasing weight gain and lipid deposition in the liver and adipose tissue | [97] |
| In vitro | RAW 264.7 macrophage cells | 25 µM | Enhancing the catecholamine production, accompanying by suppressing the pro-inflammatory M1 macrophages, and activating anti-inflammatory M2 macrophages in white adipose tissue | [98] |
| In vivo | Mice | 0.2% diet | Promoting white adipose browning and thermogenesis in the male descendants, and these health benefits persisted and prevented obesity in their future life | [99] |
| In vitro; In vivo | L6 myogenic cell line; Rats | 1, 5, 10, 25 or 50 µM; 0.4% diet | In vitro: Improving mitochondrial function and reducing oxidative stress through the PKA/LKB1/AMPK pathway; In vivo: Preventing muscle loss and myofiber size decrease, improving grip strength, and abolishing excessive fat accumulation | [100] |
| In vivo | Mice | 0.06% diet | Improving obesity-related complications by restoring plasma thyroid hormone levels, and attenuating oxidative stress in the heart | [101] |
| In vitro | Human sperm | 2.6, 6, 15, 30, 50, 100 µmol/L | Improving obesity-related complications by restoring reproductive dysfunction like infertility | [102] |
| *Alzheimer's disease and Parkinson's disease* | | | | |
| In vivo | Rats | 20 mg/kg/day | Ameliorating ERS by downregulating the gene expression of CHOP and GRP78, inhibiting caspase-3 activity, and ameliorating oxidative damage via suppressing xanthine oxidase activity and protein carbonyl formation as well as activating glutathione peroxidase and Nrf2 signaling pathway | [10] |

**Table 2.** *Cont.*

| Study Type | Subject | Dose | Main Findings | Ref. |
|---|---|---|---|---|
| In vitro | CL2006 cells | 100 μM | Inhibiting the aggregation of Aβ by modulating specific proteins such as UBL/XBP-1 involved in proteostasis | [103] |
| In vivo | Mice | 16 mg/kg/day | Preventing memory loss by decreasing elevated levels of mitochondrial complex IV protein in the mouse brain via the activation of SIRT1 and AMPK pathways | [104] |
| In vivo | Mice | 100 mg/kg/day | Preventing memory loss via the activation of SIRT1 and AMPK pathways | [105] |
| In vitro; In vivo | SH-SY5Y cells; Mice | 50 μM; 50 mg/kg | Elevating miR-214 expression, leading to decreased mRNA expression of α-synuclein | [106] |
| *Sex-dependent effects of resveratrol* | | | | |
| In vivo | Rats | 2.5 mg/kg/day | Superior improvements of MI in females in terms of IVSDs, ESV, EF, FS, and IVRT, among which IVRT is purely sex-dependent | [109] |
| In vivo | Rats | 50 mg/L in drinking water | Increasing the relaxations to estrogen in aortae, more potent in males, probably due to resveratrol's promoting nitric oxide and/or suppressing superoxide effects | [110] |
| In vitro; In vivo | MESC2.10 and SN4741 cells; Mice | 20 mg/kg; 10 μM | Increasing DAT in the striatum in females but not in males; Upregulating DAT in the dopaminergic cells by inducing its gene transcription | [111] |
| In vivo | Mice | 100 mg/kg | Adverse effects in females but not in males, regarding weight loss, stool consistency, and discomfort | [112] |

Abbreviations used in the table: AC, acetyl; AhR, aryl hydrocarbon receptor; Akt, protein Kinase B; AMPK, AMP-activated protein kinase; Aβ, amyloid β; cAMP, cyclic adenosine monophosphate; CAT, catalase; cGMP, cyclic guanosine monophosphate; CHOP, C/EBP homologous protein; COX-2, cyclooxygenase-2; DAG, diacylglycerol; DAT, dopamine transporter; EF, ejection fraction; EGFR, epidermal growth factor receptor; eNOS, endothelial nitric oxide synthase; ERK, extracellular signal-regulated kinases; ERRα, estrogen related receptor α; ERS, endoplasmic reticulum stress; Erα, estrogen receptor α; ERα+, estrogen receptor alpha positive; ESV, end systolic volume; EV71, enterovirus 71; FERM, band 4.1, ezrin, radixin, and moesin; FoxO1, forkhead box protein O1; FS, fractional shortening; GPx, glutathione peroxidase; GRP78, glucose-regulated protein 78; GβL, G protein beta subunit-like; HFD, high-fat diet; HIF-1α, hypoxia-inducible factor 1α; HMGB1, high mobility group box 1; HMGB1, high mobility group box 1; HO-1, heme oxygenase (decycling) 1; HSL, hormone-sensitive lipase; ICAM-1, intercellular adhesion molecule-1; IFN-γ, interferon γ; IKK, IκB kinase; IL-1β, interleukin-1β; IVRT, isovolumic relaxation time; IVSDs, interventricular septal wall dimension at systole; IκBα, nuclear factor of kappa light polypeptide gene enhancer in B-cells inhibitor α; JAK, Janus kinase; Keap1, Kelch-like ECH-associated protein 1; LDLR, low-density lipoprotein receptor; LKB1, liver kinase B1; LPS, lipopolysaccharides; MALAT1, metastasis-associated lung adenocarcinoma transcript 1; MAP2K, mitogen-activated protein kinase kinase; MAPK, mitogen-activated protein kinase; MCP-1, monocyte chemoattractant protein-1; MI, myocardial infarction; mSIN1, mammalian stress-activated protein kinase interacting protein 1; mTOR, mammalian target of rapamycin; mTORC2, mTOR Complex 2; NAD, nicotinamide adenine dinucleotide; NAFLD, non-alcoholic fatty liver disease; NALP3, NACHT, LRR, and PYD domains-containing protein 3; NF-κB, nuclear factor kappa-light-chain-enhancer of activated B cells; Nrf2, nuclear factor (erythroid-derived 2)-like 2; p53, phosphoprotein p53; PDE 3B, phosphodiesterase 3B expression; PDK1, phosphoinositide dependent kinase 1; PGC, peroxisome proliferator-activated receptor gamma coactivator 1α; PI3K, phosphatidylinositol 3-kinase; PIP2, phosphatidylinositol 4,5-bisphosphate; PIP3, phosphatidylinositol-3,4,5–trisphosphate; PKA, protein kinase A; PKC-θ, protein kinase C θ; PKG1α, cGMP-dependent protein kinase 1α; PLSCR1, phospholipid scramblase 1; PPAR-γ, peroxisome proliferator-activated receptor γ; PTEN, phosphatase and tensin homolog; RANTES, regulated on activation normal T cell expressed and secreted; RICTOR, the rapamycin-insensitive companion of mTOR; SARM, sterile α and armadillo motif protein; SIRT, sirtuin 1; α-SMA, smooth muscle actin; SOD, superoxide dismutase; SRB1, scavenger receptor class B type I; STAT, signal transducer and activator of transcription; tBregs, tumor-evoked regulatory B cells; TF, transcription factor; TGF-β, transforming growth factor β; TLR4, toll-like receptor 4; TNF-α, tumor necrosis factor α; TRIF, toll/IL-1 receptor domain-containing adaptor inducing β interferon; UBL, ubiquitin-like protein; XBP-1, X-box binding protein 1.

## 4. Clinical Trials

In light of the positive epidemiological evidence and the promising results from experimental studies, resveratrol has been investigated in the human population as a potential nutraceutical (Table 3). There are certain encouraging outcomes reported. Specifically, resveratrol intake (500 mg/day for 30 days) was demonstrated to reduce CVD risk factors by increasing SIRT1, enhancing total antioxidant capacity in healthy individuals, decreasing low-density lipoprotein cholesterol (LDL-C), ApoB, and oxidized LDL [113–115]. Moreover, resveratrol prevented bone density loss (500 mg/day for 6 months) in type-2 diabetic patients [116]. In addition, resveratrol showed benefits in obesity, NAFLD, and neurodegenerative diseases [15,16,117,118]. However, some null outcomes have also been reported. For instance, resveratrol intake (250 mg/day for 8 weeks) did not increase SIRT1 nor improve many cardiovascular risk factors in healthy aged men [58]. In some other studies, no significant improvements were found in metabolic biomarkers in patients with Alzheimer's disease, obesity or type-2 diabetes, respectively, though their resveratrol intake ranged from 150 to 1000 mg/day with different duration of 4–52 weeks [119–122]. Therefore, the outcomes of clinical studies are not always consistent. Of note, the health effects of resveratrol as a therapeutic intervention may be affected by many factors, such as baseline health status of the subjects, their demographic profile, lifestyle, eating pattern, resveratrol dose, and intervention period. Nevertheless, a well-designed study, proper sample size, and a scientific evaluation system are also needed. Furthermore, although resveratrol is well-tolerated and safe as reported by most of the clinical trials, very few adverse effects (e.g., nausea and diarrhea) were observed, as well as some unfavorable results like an increase in total cholesterol, ApoB, the homeostatic model assessment-insulin resistance (HOMA-IR) score, fasting blood glucose, body fat, and the inflammatory markers [15,114,118]. Interestingly, resveratrol was reported to mask the exercise training-induced benefits, blunting the improved cardiovascular health parameters [58]. It might be attributable to the potent antioxidant capability of resveratrol, which could scavenge the free radicals induced by exercise training, because the appropriate number of free radicals is necessary for health maintenance. Therefore, it could be suggested that foods containing resveratrol should not be consumed during exercise.

**Table 3.** The results of resveratrol from clinical research.

| Population | Targeting Diseases | Study Type | Sample Size (Valid Data) | Resveratrol Dose and Duration | Main Findings: Resveratrol vs. Measurements/Risk Factors/Biomarkers | Ref. |
|---|---|---|---|---|---|---|
| Healthy and slightly overweight | CVD—atherosclerosis | Randomized, parallel | $N = 48$ (male, 24; female, 24) | Resveratrol supplement, 500 mg/day (30 days) | *Favorable:* Increased serum SIRT1 concentrations from $1.06 \pm 0.71$ to $5.75 \pm 2.98$ ng/mL, $p < 0.0001$<br>*Null:* Did not influence the various metabolic parameters (BW, BMI, WC, HR, BP, HDL, LDL, TG, BG, estradiol, estrone, insulin, hsCRP, and TAC)<br>Unfavorable: Increased TC, ApoB, and HOMA-IR score | [114] |
| Asymptomatic hypercholesterolemics (AHCs) and normohypercholemics (NC) | CVD—atherogenesis | Randomized, placebo-controlled | $N = 40$ (male, 21; female, 19) | Resveratrol supplement, 150 mg/day (4 weeks) | *Favorable:* Increased TAC (mean value increased to 136.7% after consumption, $p = 0.035$) in healthy NC individuals and facilitated an increase in vitamin E (7.18 µmol/l, i.e., 35.72%) in AHC<br>*Null:* No differences found in TC, TG, HDL, LDL, TAC (in AHC), and vitamin E (in NC) | [113] |
| Overweight and slightly obese volunteers | CVD—endothelial function | Randomized, double-blind, placebo-controlled | $N = 45$ (male, 25; female, 20) | Trans-resveratrol supplement, 150 mg/day (4 weeks) | *Null:* Did not improve endothelial function (FMD, arterial stiffness, and other endothelial activation markers), inflammation (IL-6 and TNF-α), glucose and lipid metabolism (BG, insulin, and serum TG) | [122] |
| 65 years or older with peripheral artery disease (PAD) | CVD—PAD | Randomized, double-blind, placebo-controlled | $N = 66$ (male, 45; female, 21) | Trans-resveratrol supplement, 125 and 500 mg/day (6 months) | *Favorable:* 125 mg/day improved the outcome of 6-min walk test results statistically significant (95% CI: −5.7 to 39.5) but not clinically meaningful<br>*Null:* 500 mg/day showed no significant improvement | [123] |
| Patients in primary cardiovascular disease prevention | CVD—atherogenesis | Triple-blind, randomized, placebo-controlled | $N = 75$ (male, 34; female, 41) | Resveratrol-enriched grape extract, 350 mg/day (6 months) | *Favorable:* Decreased LDL-C (−4.5%, $p = 0.04$), ApoB (−9.8%, $p = 0.014$), oxidized LDL (−20%, $p = 0.001$), and oxidized LDL/ApoB (−12.5%, $p = 0.000$); increased ratio non-HDL-C/ApoB (8.5%, $p = 0.046$)<br>*Null:* No clinically significant effects on hepatic, thyroid, and renal function (GGT, AST, ALP, bilirubin and albumin; TSH, T4; CPK, creatinine, and urate) | [115] |
| Healthy aged men | CVD | Randomized, double-blind, placebo-controlled | $N = 27$ (male) | Trans-resveratrol supplement, 250 mg/day (8 weeks) | *Null:* No effects on SIRT1 protein concentrations or cardiovascular parameters (BG, TC, and HDL), and VCAM-1<br>*Unfavorable:* Abolished the exercise training-induced improvement in maximal oxygen uptake, BP, and lipids (LDL, TC/HDL ratio, and TG) | [58] |
| Women at increased breast cancer risk | Cancer—breast cancer | Randomized, double-blind, placebo-controlled | $N = 39$ (male) | Trans-resveratrol supplement, 10 or 100 mg/day (12 weeks) | *Favorable:* Decreased the methylation of RASSF-1α ($p = 0.047$)<br>*Null:* Did not significantly alter PGE2 | [124] |
| Patients with type-2 diabetes | Type 2 diabetes | Randomized, double-blind, placebo-controlled | $N = 192$ (male, 126; female, 66) | Resveratrol supplement, 40 and 500 mg/day (6 months) | *Favorable:* Prevented bone density loss (500 mg/d) (whole-body BMD (0.01 vs. −0.03 g/cm$^2$, $p = 0.001$), whole-body BMC (4.04 vs. −58.8 g, $p < 0.001$), whole-body T-score (0.15 vs. −0.26), and serum phosphorus (0.07 vs. −0.01 µmol/L, $p = 0.002$)); decreased CRP (not significantly)<br>*Null:* BW, BMI, WC, BP, FBG, HbA1c, insulin, HOMA-IR, C-peptide, FFAs, ALT, AST, GGT, uric acid, IL-6, and adiponectin<br>*Unfavorable:* Slightly increased TC and TG (500 mg/d) | [116] |
| Patients with diet-controlled type-2 diabetes | Type 2 diabetes | Randomized, double-blind, placebo-controlled | $N = 14$ (male) | Resveratrol capsules, 1000 mg/day (5 weeks) | *Favorable:* Modestly decreased FBG and HbA1c<br>*Null:* No significant effects on GLP-1 secretion, gastric emptying, glycemic control (HbA1c, BG), energy intake, and BW | [121] |

**Table 3.** *Cont.*

| Population | Targeting Diseases | Study Type | Sample Size (Valid Data) | Resveratrol Dose and Duration | Main Findings: Resveratrol vs. Measurements/Risk Factors/Biomarkers | Ref. |
|---|---|---|---|---|---|---|
| Obese men | Obesity | Randomized, placebo-controlled | N = 24 (male) | Trans-resveratrol tablets, 500 mg/day (4 weeks) | *Null:* Did not improve insulin sensitivity, BP, BG, insulin, HOMA-IR, HbA1c, lipids (TC, HDL-C, LDL-C and TG), liver and skeletal muscle lipid content, or inflammatory biomarkers (IL-6, TNF-$\alpha$, and MCP1) and some metabolic markers<br>Unfavorable: Insignificantly deteriorated insulin sensitivity | [120] |
| Overweight/obese with insulin-resistance subjects | Obesity | Randomized, double-blind, placebo-controlled | N = 108 (male, 54; female, 54) | Resveratrol supplement, 150 mg/day (12 weeks) | *Null:* Did not significantly impact liver fat content or cardiometabolic risk biomarkers (FBG, HbA1c, BP, TC, HDL, LDL, TG, AST, ALT, GGT, hsCRP, and IL-6) | [119] |
| Overweight/obese with NAFLD | Obesity—NAFLD | Randomized, placebo-controlled | N = 75 (male, 52; female, 23) | Resveratrol capsules, 600 mg/day (12 weeks) | *Favorable:* Significantly reduced BW (95% CI: −1.61 to −0.38) and BMI (95% CI: −0.54 to −0.12)<br>*Null:* Did not significantly change ALT, AST, and lipid profiles (TG, TC, LDL-C, HDL-C), hepatic steatosis grade (ALT and AST), serum glycemic parameters (FBG, insulin, and HbA1c), and SIRT1 levels | [117] |
| Obese men | Obesity—bone health | Randomized, double-blind, placebo-controlled | N = 66 (male) | Trans-resveratrol tablets, 1000 or 150 mg/day, (16 weeks) | *Favorable:* Dose-dependently benefited bone by stimulating formation or mineralization as significantly increased BAP (R = 0.471, $p < 0.001$), and BMD (BAP and lumbar spine volumetric BMD were positively correlated: R = 0.281, $p = 0.027$) | [16] |
| Individuals with mild/moderate Alzheimer disease (AD) | Aging—AD | Randomized, double-blind, placebo-controlled | N = 119 (male, 51; female, 68) | Resveratrol supplement, 500–2000 mg/day (52 weeks) | *Uncertain:* Resveratrol and its major metabolites were detectable in plasma and CSF, suggesting CNS effects; CSF A$\beta$40 and plasma A$\beta$40 levels declined less than those in the placebo group; brain volume loss was more compared to placebo<br>*Null:* No effects on plasma A$\beta$42, CSF A$\beta$42, CSF tau, CSF phospho-tau 181, hippocampal volume, entorhinal cortex thickness, MMSE, CDR, ADAS-cog, NPI, or glucose and insulin metabolism<br>*Unfavorable:* The most common adverse events were nausea and diarrhea, but similar to placebo | [118] |
| Elderly participants | Aging—memory | Randomized, double-blind, placebo-controlled | N = 53 (male, 25; female, 28) | Resveratrol pills, 200 mg/day, (26 weeks) | *Favorable:* Non-significant trend for stable performance in a pattern recognition task<br>*Null:* No significant changes in CVLT performance, HbA1c levels, hippocampus volume, microstructure, and functional connectivity<br>*Unfavorable:* Increased serum cholesterol, weight, body fat, FBG, and inflammatory markers; decreased physical activity and neurotrophic factors<br>*Adverse events:* Two dropouts (a sudden decrease in eyesight and a skin rash), and others (case no.): Diarrhea ($n = 3$), skin changes ($n = 3$), stomach aches ($n = 1$), dizziness ($n = 1$), improved mood changes ($n = 1$), loss of hair ($n = 1$) | [15] |

Abbreviations used in the table: ADAS-cog, Alzheimer's Disease Assessment Scale–cognitive; AHC, asymptomatic hypercholesterolemics; ALP, alkaline phosphatase; ALT, aminotransferase; ApoB, apolipoprotein B; AST, aminotransferase; A$\beta$40, amyloid $\beta$40; BAP, bone alkaline phosphatase; BG, blood glucose; BMC, bone mineral content; BMD, bone mineral density; BMI, body mass index; BP, blood pressure; BW, body weight; CDR, clinical dementia rating; CNS, central nervous system; CPK, creatine phosphokinase; CRP, C-reactive protein; CSF, cerebrospinal fluid; CVLT, California Verbal Learning Task; CVR, cerebrovascular responsiveness; FBG, fasting blood glucose; FFAs, free fatty acids; FMD, flow-mediated vasodilation; GGT, $\gamma$-glutamyl transferase; GLP-1, glucagon-like peptide 1; HbA1c, glycated hemoglobin; HDL, high-density lipoprotein; HOMA-IR, the homeostatic model assessment—insulin resistance; HR, heart rate; hsCRP, high-sensitivity C-reactive protein; IL-6, interleukin-6; LDL-C, low-density lipoprotein cholesterol; MCP1, monocyte chemoattractant protein 1; MMSE, mini-mental state examination; NAFLD, non-alcoholic fatty liver disease; NC, normohypercholemics; NPI, neuropsychiatric inventory; PAD, peripheral artery disease; PGE2, prostaglandin E2; SIRT1, sirtuin 1; T4, thyroxine; TAC, total antioxidant capacity; TC, total cholesterol; TG, triglycerides; TNF-$\alpha$, tumor necrosis factor $\alpha$; TSH, thyroid-stimulating hormone; VCAM-1, vascular cell adhesion molecule 1; WC, waist circumference; WHR, waist-hip ratio.

## 5. Conclusions

Resveratrol is one of the most investigated bioactive compounds in foods. A number of epidemiologic studies have demonstrated that resveratrol is effective in the prevention of some diseases such as CVDs and cancer, although the results are sometimes inconsistent. In addition, the experimental studies have shown that resveratrol possesses many bioactivities and health benefits like antioxidant, anti-inflammatory, immunomodulatory effects, and improving CVDs, cancer, liver diseases, diabetes, obesity, Alzheimer's disease, and Parkinson's disease. Furthermore, resveratrol showed some effects in patients with CVDs and obesity in clinical trials, although inconsistency has also been reported. In the future, more bioactivities and health benefits of resveratrol should be evaluated, and further clarification of the underlying mechanisms of action is required. In order to develop resveratrol into functional foods and pharmaceuticals, more clinical trials are essential to confirm its efficacy and observe the possible adverse events, and the dose–effect relationship should be paid special attention as well.

**Author Contributions:** Conceptualization, X.M., R.-Y.G., and H.-B.L.; Writing—original draft preparation, X.M. and J.Z.; Writing—review and editing, C.-N.Z., R.-Y.G., and H.-B.L.; supervision, R.-Y.G. and H.-B.L.; funding acquisition, R.-Y.G., and H.-B.L. All authors have read and agreed to the published version of the manuscript.

**Funding:** This research was funded by China Central Public-interest Scientific Institution Basal Research Fund (Grant No. Y2020XK05), the National Key R&D Program of China (Grant No. 2018YFC1604400), and the Key Project of Guangdong Provincial Science and Technology Program (Grant No. 2014B020205002).

**Conflicts of Interest:** The authors declare no conflict of interest.

## References

1. Meng, X.; Li, Y.; Li, S.; Zhou, Y.; Gan, R.-Y.; Xu, D.-P.; Li, H.-B. Dietary sources and bioactivities of melatonin. *Nutrients* **2017**, *9*, 367. [CrossRef] [PubMed]
2. Zhao, C.-N.; Meng, X.; Li, Y.; Li, S.; Liu, Q.; Tang, G.-Y.; Li, H.-B. Fruits for prevention and treatment of cardiovascular diseases. *Nutrients* **2017**, *9*, 598. [CrossRef] [PubMed]
3. Weiskirchen, S.; Weiskirchen, R. Resveratrol: How much wine do you have to drink to stay healthy? *Adv. Nutr.* **2016**, *7*, 706–718. [CrossRef] [PubMed]
4. Jang, M.; Cai, L.; Udeani, G.O.; Slowing, K.V.; Thomas, C.F.; Beecher, C.W.; Fong, H.H.; Farnsworth, N.R.; Kinghorn, A.D.; Mehta, R.G.; et al. Cancer chemopreventive activity of resveratrol, a natural product derived from grapes. *Science* **1997**, *275*, 218–220. [CrossRef] [PubMed]
5. Lin, Y.; Yngve, A.; Lagergren, J.; Lu, Y. A dietary pattern rich in lignans, quercetin and resveratrol decreases the risk of oesophageal cancer. *Br. J. Nutr.* **2014**, *112*, 2002–2009. [CrossRef] [PubMed]
6. Zamora-Ros, R.; Urpi-Sarda, M.; Lamuela-Raventós, R.M.; Martínez-González, M.Á.; Salas-Salvadó, J.; Arós, F.; Fitó, M.; Lapetra, J.; Estruch, R.; Andres-Lacueva, C. High urinary levels of resveratrol metabolites are associated with a reduction in the prevalence of cardiovascular risk factors in high-risk patients. *Pharmacol. Res.* **2012**, *65*, 615–620. [CrossRef]
7. Mattison, J.A.; Wang, M.; Bernier, M.; Zhang, J.; Park, S.; Maudsley, S.; An, S.S.; Santhanam, L.; Martin, B.; Faulkner, S.; et al. Resveratrol prevents high fat/sucrose diet-induced central arterial wall inflammation and stiffening in nonhuman primates. *Cell Metab.* **2014**, *20*, 183–190. [CrossRef]
8. Chong, E.; Chang, S.; Hsiao, Y.; Singhal, R.; Liu, S.; Leha, T.; Lin, W.; Hsu, C.; Chen, Y.; Chen, Y.; et al. Resveratrol, a red wine antioxidant, reduces atrial fibrillation susceptibility in the failing heart by PI3K/AKT/eNOS signaling pathway activation. *Heart Rhythm* **2015**, *12*, 1046–1056. [CrossRef]
9. Nishikawa, K.; Iwaya, K.; Kinoshita, M.; Fujiwara, Y.; Akao, M.; Sonoda, M.; Thiruppathi, S.; Suzuki, T.; Hiroi, S.; Seki, S.; et al. Resveratrol increases CD68[+] Kupffer cells colocalized with adipose differentiation-related protein and ameliorates high-fat-diet-induced fatty liver in mice. *Mol. Nutr. Food Res.* **2015**, *59*, 1155–1170. [CrossRef]
10. Gaballah, H.H.; Zakaria, S.S.; Elbatsh, M.M.; Tahoon, N.M. Modulatory effects of resveratrol on endoplasmic reticulum stress-associated apoptosis and oxido-inflammatory markers in a rat model of rotenone-induced Parkinson's disease. *Chem. Biol. Interact.* **2016**, *251*, 10–16. [CrossRef]

11. Wu, H.; Sheng, Z.; Xie, J.; Li, R.; Chen, L.; Li, G.; Wang, L.; Xu, B. Reduced HMGB 1-mediated pathway and oxidative stress in resveratrol-treated diabetic mice: A possible mechanism of cardioprotection of resveratrol in diabetes mellitus. *Oxid. Med. Cell. Longev.* **2016**, *2016*. [CrossRef] [PubMed]

12. Moussa, C.; Hebron, M.; Huang, X.; Ahn, J.; Rissman, R.A.; Aisen, P.S.; Turner, R.S. Resveratrol regulates neuro-inflammation and induces adaptive immunity in Alzheimer's disease. *J. Neuroinflamm.* **2017**, *14*, 1. [CrossRef] [PubMed]

13. Sung, M.M.; Kim, T.T.; Denou, E.; Soltys, C.M.; Hamza, S.M.; Byrne, N.J.; Masson, G.; Park, H.; Wishart, D.S.; Madsen, K.L.; et al. Improved glucose homeostasis in obese mice treated with resveratrol is associated with alterations in the gut microbiome. *Diabetes* **2017**, *66*, 418–425. [CrossRef] [PubMed]

14. Cheng, T.M.; Chin, Y.T.; Ho, Y.; Chen, Y.R.; Yang, Y.N.; Yang, Y.C.; Shih, Y.J.; Lin, T.I.; Lin, H.Y.; Davis, P.J. Resveratrol induces sumoylated COX-2-dependent anti-proliferation in human prostate cancer LNCaP cells. *Food Chem. Toxicol.* **2018**, *112*, 67–75. [CrossRef] [PubMed]

15. Huhn, S.; Beyer, F.; Zhang, R.; Lampe, L.; Grothe, J.; Kratzsch, J.; Willenberg, A.; Breitfeld, J.; Kovacs, P.; Stumvoll, M.; et al. Effects of resveratrol on memory performance, hippocampus connectivity and microstructure in older adults—A randomized controlled trial. *Neuroimage* **2018**, *174*, 177–190. [CrossRef] [PubMed]

16. Ornstrup, M.J.; Harslof, T.; Kjaer, T.N.; Langdahl, B.L.; Pedersen, S.B. Resveratrol increases bone mineral density and bone alkaline phosphatase in obese men: A randomized placebo-controlled trial. *J. Clin. Endocr. Metab.* **2014**, *99*, 4720–4729. [CrossRef]

17. Levi, F.; Pasche, C.; Lucchini, F.; Ghidoni, R.; Ferraroni, M.; La Vecchia, C. Resveratrol and breast cancer risk. *Eur. J. Cancer Prev.* **2005**, *14*, 139–142. [CrossRef]

18. Rabassa, M.; Zamora-Ros, R.; Urpi-Sarda, M.; Bandinelli, S.; Ferrucci, L.; Andres-Lacueva, C.; Cherubini, A. Association of habitual dietary resveratrol exposure with the development of frailty in older age: The Invecchiare in Chianti study. *Am. J. Clin. Nutr.* **2015**, *102*, 1534–1542. [CrossRef]

19. Sohrab, G.; Hosseinpour-Niazi, S.; Hejazi, J.; Yuzbashian, E.; Mirmiran, P.; Azizi, F. Dietary polyphenols and metabolic syndrome among Iranian adults. *Int. J. Food Sci. Nutr.* **2013**, *64*, 661–667. [CrossRef]

20. Renaud, S.; de Lorgeril, M. Wine, alcohol, platelets, and the French paradox for coronary heart disease. *Lancet* **1992**, *339*, 1523–1526. [CrossRef]

21. Tresserra-Rimbau, A.; Rimm, E.B.; Medina-Remon, A.; Martinez-Gonzalez, M.A.; de la Torre, R.; Corella, D.; Salas-Salvado, J.; Gomez-Gracia, E.; Lapetra, J.; Aros, F.; et al. Inverse association between habitual polyphenol intake and incidence of cardiovascular events in the PREDIMED study. *Nutr. Metab. Carbiovasc. Dis.* **2014**, *24*, 639–647. [CrossRef] [PubMed]

22. Tresserra-Rimbau, A.; Rimm, E.B.; Medina-Remon, A.; Martinez-Gonzalez, M.A.; Lopez-Sabater, M.C.; Covas, M.I.; Corella, D.; Salas-Salvado, J.; Gomez-Gracia, E.; Lapetra, J.; et al. Polyphenol intake and mortality risk: A re-analysis of the PREDIMED trial. *BMC Med.* **2014**, *12*, 77. [CrossRef] [PubMed]

23. Li, G.P.; Zhu, Y.P.; Zhang, Y.P.; Lang, J.M.; Chen, Y.P.; Ling, W.P. Estimated daily flavonoid and stilbene intake from fruits, vegetables, and nuts and associations with lipid profiles in Chinese adults. *J. Acad. Nutr. Diet.* **2013**, *113*, 786–794. [CrossRef] [PubMed]

24. Pan, Y.; Zhang, H.; Zheng, Y.; Zhou, J.; Yuan, J.; Yu, Y.; Wang, J. Resveratrol exerts antioxidant effects by activating SIRT2 to deacetylate Prx1. *Biochemistry US* **2017**, *56*, 6325–6328. [CrossRef] [PubMed]

25. Yun, H.; Park, S.; Kim, M.; Yang, W.K.; Im, D.U.; Yang, K.R.; Hong, J.; Choe, W.; Kang, I.; Kim, S.S.; et al. AMP-activated protein kinase mediates the antioxidant effects of resveratrol through regulation of the transcription factor FoxO1. *FEBS J.* **2014**, *281*, 4421–4438. [CrossRef] [PubMed]

26. Ingles, M.; Gambini, J.; Miguel, M.G.; Bonet-Costa, V.; Abdelaziz, K.M.; El Alami, M.; Vina, J.; Borras, C. PTEN mediates the antioxidant effect of resveratrol at nutritionally relevant concentrations. *Biomed. Res. Int.* **2014**, 580852. [CrossRef]

27. Singh, A.K.; Vinayak, M. Resveratrol alleviates inflammatory hyperalgesia by modulation of reactive oxygen species (ROS), antioxidant enzymes and ERK activation. *Inflamm. Res.* **2017**, *66*, 911–921. [CrossRef]

28. Fu, S.; Lv, R.; Wang, L.; Hou, H.; Liu, H.; Shao, S. Resveratrol, an antioxidant, protects spinal cord injury in rats by suppressing MAPK pathway. *Saudi J. Biol. Sci.* **2018**, *25*, 259–266. [CrossRef]

29. Li, J.; Feng, L.; Xing, Y.; Wang, Y.; Du, L.; Xu, C.; Cao, J.; Wang, Q.; Fan, S.; Liu, Q.; et al. Radioprotective and antioxidant effect of resveratrol in hippocampus by activating Sirt1. *Int. J. Mol. Sci.* **2014**, *15*, 5928–5939. [CrossRef]

30. Li, S.; Zlia, G.; Chen, L.; Ding, Y.; Lian, J.; Hong, G.; Lu, Z. Resveratrol protects mice from paraquat-induced lung injury: The important role of SIRT1 and NRF2 antioxidant pathways. *Mol. Med. Rep.* **2016**, *13*, 1833–1838. [CrossRef]

31. Meng, Q.; Guo, T.; Li, G.; Sun, S.; He, S.; Cheng, B.; Shi, B.; Shan, A. Dietary resveratrol improves antioxidant status of sows and piglets and regulates antioxidant gene expression in placenta by Keap1-Nrf2 pathway and Sirt1. *J. Anim. Sci. Biotechnol.* **2018**, *9*, 34. [CrossRef] [PubMed]

32. Zhou, X.; Yang, J.; Zhou, M.; Zhang, Y.; Liu, Y.; Hou, P.; Zeng, X.; Yi, L.; Mi, M. Resveratrol attenuates endothelial oxidative injury by inducing autophagy via the activation of transcription factor EB. *Nutr. Metab.* **2019**, *16*, 42. [CrossRef] [PubMed]

33. Kim, J.; Kundu, M.; Viollet, B.; Guan, K. AMPK and mTOR regulate autophagy through direct phosphorylation of Ulk1. *Nat. Cell Biol.* **2011**, *13*, 132–141. [CrossRef] [PubMed]

34. Yang, S.J.; Lim, Y. Resveratrol ameliorates hepatic metaflammation and inhibits NLRP3 inflammasome activation. *Metabolism* **2014**, *63*, 693–701. [CrossRef] [PubMed]

35. Misawa, T.; Saitoh, T.; Kozaki, T.; Park, S.; Takahama, M.; Akira, S. Resveratrol inhibits the acetylated alpha-tubulin-mediated assembly of the NLRP3-inflammasome. *Int. Immunol.* **2015**, *27*, 425–434. [CrossRef] [PubMed]

36. Andrews, C.S.; Matsuyama, S.; Lee, B.; Li, J. Resveratrol suppresses NTHi-induced inflammation via upregulation of the negative regulator MyD88 short. *Sci. Rep.* **2016**, *6*, 34445. [CrossRef] [PubMed]

37. Chen, Y.H.; Fu, Y.C.; Wu, M.J. Does resveratrol play a role in decreasing the inflammation associated with contrast induced nephropathy in rat model? *J. Clin. Med.* **2019**, *8*, 147. [CrossRef]

38. Son, Y.; Chung, H.; Pae, H. Differential effects of resveratrol and its natural analogs, piceatannol and 3,5,4′-trans-trimethoxystilbene, on anti-inflammatory heme oxigenase-1 expression in RAW264.7 macrophages. *Biofactors* **2014**, *40*, 138–145. [CrossRef]

39. Nwachukwu, J.C.; Srinivasan, S.; Bruno, N.E.; Parent, A.A.; Hughes, T.S.; Pollock, J.A.; Gjyshi, O.; Cavett, V.; Nowak, J.; Garcia-Ordonez, R.D.; et al. Resveratrol modulates the inflammatory response via an estrogen receptor-signal integration network. *ELife* **2014**, *3*, e2057. [CrossRef]

40. Sadeghi, A.; Ebrahimi, S.S.S.; Golestani, A.; Meshkani, R. Resveratrol ameliorates palmitate-induced inflammation in skeletal muscle cells by attenuating oxidative stress and JNK/NF-κB pathway in a SIRT1-independent mechanism. *J. Cell. Biochem.* **2017**, *118*, 2654–2663. [CrossRef]

41. Ma, C.; Wang, Y.; Dong, L.; Li, M.; Cai, W. Anti-inflammatory effect of resveratrol through the suppression of NF-κB and JAK/STAT signaling pathways. *Acta Bioch. Bioph. Sin.* **2015**, *47*, 207–213. [CrossRef] [PubMed]

42. Pinheiro, D.; de Oliveira, A.; Coutinho, L.G.; Fontes, F.L.; de Medeiros, O.R.; Oliveira, T.T.; Faustino, A.; Lira, D.S.V.; de Melo, C.J.; Lajus, T.; et al. Resveratrol decreases the expression of genes involved in inflammation through transcriptional regulation. *Free Radic. Biol. Med.* **2019**, *130*, 8–22. [CrossRef] [PubMed]

43. Li, J.; Li, L.; Wang, S.; Zhang, C.; Zheng, L.; Jia, Y.; Xu, M.; Zhu, T.; Zhang, Y.; Rong, R. Resveratrol alleviates inflammatory responses and oxidative stress in rat kidney ischemia-reperfusion injury and $H_2O_2$-induced NRK-52E Cells via the Nrf2/TLR4/NF-κB Pathway. *Cell. Physiol. Biochem.* **2018**, *45*, 1677–1689. [CrossRef] [PubMed]

44. Wang, G.; Hu, Z.; Fu, Q.; Song, X.; Cui, Q.; Jia, R.; Zou, Y.; He, C.; Li, L.; Yin, Z. Resveratrol mitigates lipopolysaccharide-mediated acute inflammation in rats by inhibiting the TLR4/NF-κBp65/MAPKs signaling cascade. *Sci. Rep.* **2017**, *7*, 45006. [CrossRef] [PubMed]

45. Limagne, E.; Lancon, A.; Delmas, D.; Cherkaoui-Malki, M.; Latruffe, N. Resveratrol interferes with IL1β-induced pro-inflammatory paracrine interaction between primary chondrocytes and macrophages. *Nutrients* **2016**, *8*, 280. [CrossRef] [PubMed]

46. Euba, B.; Lopez-Lopez, N.; Rodriguez-Arce, I.; Fernandez-Calvet, A.; Barberan, M.; Caturla, N.; Marti, S.; Diez-Martinez, R.; Garmendia, J. Resveratrol therapeutics combines both antimicrobial and immunomodulatory properties against respiratory infection by nontypeable *Haemophilus influenzae*. *Sci. Rep.* **2017**, *7*, 12860. [CrossRef]

47. Mastromarino, P.; Capobianco, D.; Cannata, F.; Nardis, C.; Mattia, E.; De Leo, A.; Restignoli, R.; Francioso, A.; Mosca, L. Resveratrol inhibits rhinovirus replication and expression of inflammatory mediators in nasal epithelia. *Antivir. Res.* **2015**, *123*, 15–21. [CrossRef]

48. Zhang, L.; Li, Y.; Gu, Z.; Wang, Y.; Shi, M.; Ji, Y.; Sun, J.; Xu, X.; Zhang, L.; Jiang, J.; et al. Resveratrol inhibits enterovirus 71 replication and pro-inflammatory cytokine secretion in rhabdosarcoma cells through blocking IKKs/NF-κB signaling pathway. *PLoS ONE* **2015**, *10*, e1168792. [CrossRef]

49. Zhang, C.; Tian, Y.; Yan, F.; Kang, X.; Han, R.; Sun, G.; Zhang, H. Modulation of growth and immunity by dietary supplementation with resveratrol in young chickens receiving conventional vaccinations. *Am. J. Vet. Res.* **2014**, *75*, 752–759. [CrossRef]

50. Liu, T.; Zang, N.; Zhou, N.; Li, W.; Xie, X.; Deng, Y.; Ren, L.; Long, X.; Li, S.; Zhou, L.; et al. Resveratrol inhibits the TRIF-dependent pathway by upregulating sterile α and armadillo motif protein, contributing to anti-inflammatory effects after respiratory syncytial virus infection. *J. Virol.* **2014**, *88*, 4229–4236. [CrossRef]

51. Cui, Q.; Fu, Q.; Zhao, X.; Song, X.; Yu, J.; Yang, Y.; Sun, K.; Bai, L.; Tian, Y.; Chen, S.; et al. Protective effects and immunomodulation on piglets infected with rotavirus following resveratrol supplementation. *PLoS ONE* **2018**, *13*, e1926922. [CrossRef] [PubMed]

52. Smith, N.C.; Christian, S.L.; Taylor, R.G.; Santander, J.; Rise, M.L. Immune modulatory properties of 6-gingerol and resveratrol in Atlantic salmon macrophages. *Mol. Immunol.* **2018**, *95*, 10–19. [CrossRef] [PubMed]

53. Huang, T.; Chen, C.; Liu, H.; Lee, T.; Shieh, S. Resveratrol pretreatment attenuates concanavalin A-induced hepatitis through reverse of aberration in the immune response and regenerative capacity in aged mice. *Sci. Rep.* **2017**, *7*, 2705. [CrossRef] [PubMed]

54. Lai, X.; Pei, Q.; Song, X.; Zhou, X.; Yin, Z.; Jia, R.; Zou, Y.; Li, L.; Yue, G.; Liang, X.; et al. The enhancement of immune function and activation of NF-κB by resveratrol-treatment in immunosuppressive mice. *Int. Immunopharmacol.* **2016**, *33*, 42–47. [CrossRef]

55. Warburton, A.; Vasieva, O.; Quinn, P.; Stewart, J.P.; Quinn, J.P. Statistical analysis of human microarray data shows that dietary intervention with n-3 fatty acids, flavonoids and resveratrol enriches for immune response and disease pathways. *Br. J. Nutr.* **2018**, *119*, 239–249. [CrossRef]

56. Guo, N.H.; Fu, X.; Zi, F.M.; Song, Y.; Wang, S.; Cheng, J. The potential therapeutic benefit of resveratrol on Th17/Treg imbalance in immune thrombocytopenic purpura. *Int. Immunopharmacol.* **2019**, *73*, 181–192. [CrossRef]

57. Buttari, B.; Profumo, E.; Segoni, L.; D'Arcangelo, D.; Rossi, S.; Facchiano, F.; Saso, L.; Businaro, R.; Iuliano, L.; Rigano, R. Resveratrol counteracts inflammation in human M1 and M2 macrophages upon challenge with 7-oxo-cholesterol: Potential therapeutic implications in atherosclerosis. *Oxid. Med. Cell. Longev.* **2014**, 257543. [CrossRef]

58. Gliemann, L.; Schmidt, J.F.; Olesen, J.; Bienso, R.S.; Peronard, S.L.; Grandjean, S.U.; Mortensen, S.P.; Nyberg, M.; Bangsbo, J.; Pilegaard, H.; et al. Resveratrol blunts the positive effects of exercise training on cardiovascular health in aged men. *J. Physiol.* **2013**, *591*, 5047–5059. [CrossRef]

59. Vasamsetti, S.B.; Karnewar, S.; Gopoju, R.; Gollavilli, P.N.; Narra, S.R.; Kumar, J.M.; Kotamraju, S. Resveratrol attenuates monocyte-to-macrophage differentiation and associated inflammation via modulation of intracellular GSH homeostasis: Relevance in atherosclerosis. *Free Radic. Biol. Med.* **2016**, *96*, 392–405. [CrossRef]

60. Javkhedkar, A.A.; Quiroz, Y.; Rodriguez-Iturbe, B.; Vaziri, N.D.; Lokhandwala, M.F.; Banday, A.A. Resveratrol restored Nrf2 function, reduced renal inflammation, and mitigated hypertension in spontaneously hypertensive rats. *Am. J. Physiol. Reg. I.* **2015**, *308*, R840–R846. [CrossRef]

61. Prysyazhna, O.; Wolhuter, K.; Switzer, C.; Santos, C.; Yang, X.; Lynham, S.; Shah, A.M.; Eaton, P.; Burgoyne, J.R. Blood pressure-lowering by the antioxidant resveratrol is counterintuitively mediated by oxidation of cGMP-dependent protein kinase. *Circulation* **2019**, *140*, 126–137. [CrossRef] [PubMed]

62. Deng, Z.Y.; Hu, M.M.; Xin, Y.F.; Gang, C. Resveratrol alleviates vascular inflammatory injury by inhibiting inflammasome activation in rats with hypercholesterolemia and vitamin $D_2$ treatment. *Inflamm. Res.* **2015**, *64*, 321–332. [CrossRef] [PubMed]

63. Seo, Y.; Park, J.; Choi, W.; Ju Son, D.; Sung Kim, Y.; Kim, M.; Yoon, B.; Pyee, J.; Tae Hong, J.; Go, Y.; et al. Antiatherogenic effect of resveratrol attributed to decreased expression of ICAM-1 (intercellular adhesion molecule-1). *Arterioscler. Thromb. Vasc. Biol.* **2019**, *39*, 675–684. [CrossRef] [PubMed]

64. Fernandez-Castillejo, S.; Macia, A.; Motilva, M.J.; Catalan, U.; Sola, R. Endothelial cells deconjugate resveratrol metabolites to free resveratrol: A possible role in tissue factor modulation. *Mol. Nutr. Food Res.* **2019**, *63*, e1800715. [CrossRef]

65. Dillenburg, D.R.; Mostarda, C.; Moraes-Silva, I.C.; Ferreira, D.; Goncalves Bos, D.D.S.; Machado Duarte, A.A.; Irigoyen, M.C.; Rigatto, K. Resveratrol and grape juice differentially ameliorate cardiovascular autonomic modulation in L-NAME-treated rats. *Auton. Neurosci. Basic* **2013**, *179*, 9–13. [CrossRef]

66. Tellone, E.; De Rosa, M.C.; Pirolli, D.; Russo, A.; Giardina, B.; Galtieri, A.; Ficarra, S. Molecular interactions of hemoglobin with resveratrol: Potential protective antioxidant role and metabolic adaptations of the erythrocyte. *Biol. Chem.* **2014**, *395*, 347–354. [CrossRef]

67. Lee-Chang, C.; Bodogai, M.; Martin-Montalvo, A.; Wejksza, K.; Sanghvi, M.; Moaddel, R.; de Cabo, R.; Biragyn, A. Inhibition of breast cancer metastasis by resveratrol-mediated inactivation of tumor-evoked regulatory B cells. *J. Immunol.* **2013**, *191*, 4141–4151. [CrossRef]

68. Saud, S.M.; Li, W.; Morris, N.L.; Matter, M.S.; Colburn, N.H.; Kim, Y.S.; Young, M.R. Resveratrol prevents tumorigenesis in mouse model of Kras activated sporadic colorectal cancer by suppressing oncogenic Kras expression. *Carcinogenesis* **2014**, *35*, 2778–2786. [CrossRef]

69. Li, W.; Ma, X.; Li, N.; Liu, H.; Dong, Q.; Zhang, J.; Yang, C.; Liu, Y.; Liang, Q.; Zhang, S.; et al. Resveratrol inhibits Hexokinases II mediated glycolysis in non-small cell lung cancer via targeting Akt signaling pathway. *Exp. Cell Res.* **2016**, *349*, 320–327. [CrossRef]

70. Mikula-Pietrasik, J.; Sosinska, P.; Ksiazek, K. Resveratrol inhibits ovarian cancer cell adhesion to peritoneal mesothelium *in vitro* by modulating the production of $\alpha5\beta1$ integrins and hyaluronic acid. *Gynecol. Oncol.* **2014**, *134*, 624–630. [CrossRef]

71. Zhao, Y.; Yuan, X.; Li, X.; Zhang, Y. Resveratrol significantly inhibits the occurrence and development of cervical cancer by regulating phospholipid scramblase 1. *J. Cell. Biochem.* **2018**, *120*, 1527–1531. [CrossRef] [PubMed]

72. Ou, X.; Chen, Y.; Cheng, X.; Zhang, X.; He, Q. Potentiation of resveratrol-induced apoptosis by matrine in human hepatoma HepG2 cells. *Oncol. Rep.* **2014**, *32*, 2803–2809. [CrossRef] [PubMed]

73. Yang, Z.; Xie, Q.; Chen, Z.; Ni, H.; Xia, L.; Zhao, Q.; Chen, Z.; Chen, P. Resveratrol suppresses the invasion and migration of human gastric cancer cells via inhibition of MALAT1-mediated epithelial-to-mesenchymal transition. *Exp. Ther. Med.* **2019**, *17*, 1569–1578. [CrossRef] [PubMed]

74. Pezzuto, J.M. Resveratrol: Twenty years of growth, development and controversy. *Biomol. Ther.* **2019**, *27*, 1–14. [CrossRef] [PubMed]

75. Basly, J.P.; Marre-Fournier, F.; Bail, J.C.L.; Habrioux, G.; Chulia, A.J. Estrogenic/antiestrogenic and scavenging properties of (E)- and (Z)- resveratrol. *Life Sci.* **2000**, *66*, 769–777. [CrossRef]

76. Nakagawa, H.; Kiyozuka, Y.; Uemura, Y.; Senzaki, H.; Shikata, N.; Hioki, K.; Tsubura, A. Resveratrol inhibits human breast cancer cell growth and may mitigate the effect of linoleic acid, a potent breast cancer cell stimulator. *J. Cancer Res. Clin. Oncol. Suppl.* **2001**, *127*, 258–264. [CrossRef]

77. Folkerd, E.; Dowsett, M. Sex hormones and breast cancer risk and prognosis. *Breast* **2013**, *22*, S38–S43. [CrossRef]

78. Yildiz, F. *Phytoestrogens in Functional Foods*, 1st ed.; CRC Pres: Boca Raton, FL, USA, 2005; pp. 210–211, ISBN 9780429113802.

79. Cai, H.; Scott, E.; Kholghi, A.; Andreadi, C.; Rufini, A.; Karmokar, A.; Britton, R.G.; Horner-Glister, E.; Greaves, P.; Jawad, D.; et al. Cancer chemoprevention: Evidence of a nonlinear dose response for the protective effects of resveratrol in humans and mice. *Sci. Transl. Med.* **2015**, *7*, 298ra117. [CrossRef]

80. Ahmad, A.; Ahmad, R. Resveratrol mitigate structural changes and hepatic stellate cell activation in N'-nitrosodimethylamine-induced liver fibrosis via restraining oxidative damage. *Chem. Biol. Interact.* **2014**, *221*, 1–12. [CrossRef]

81. Ahmad, A.; Ahmad, R. Proteomic approach to identify molecular signatures during experimental hepatic fibrosis and resveratrol supplementation. *Int. J. Biol. Macromol.* **2018**, *119*, 1218–1227. [CrossRef]

82. Tanriverdi, G.; Kaya-Dagistanli, F.; Ayla, S.; Demirci, S.; Eser, M.; Unal, Z.S.; Cengiz, M.; Oktar, H. Resveratrol can prevent $CCl_4$-induced liver injury by inhibiting Notch signaling pathway. *Histol. Histopathol.* **2016**, *31*, 769–784. [PubMed]

83. Xin, P.; Han, H.; Gao, D.; Cui, W.; Yang, X.; Ying, C.; Sun, X.; Hao, L. Alleviative effects of resveratrol on nonalcoholic fatty liver disease are associated with up regulation of hepatic low density lipoprotein receptor and scavenger receptor class B type I gene expressions in rats. *Food Chem. Toxicol.* **2013**, *52*, 12–18. [CrossRef] [PubMed]

84. Li, L.; Hai, J.; Li, Z.; Zhang, Y.; Peng, H.; Li, K.; Weng, X. Resveratrol modulates autophagy and NF-κB activity in a murine model for treating non-alcoholic fatty liver disease. *Food Chem. Toxicol.* **2014**, *63*, 166–173. [CrossRef] [PubMed]

85. Ma, Z.; Zhang, Y.; Li, Q.; Xu, M.; Bai, J.; Wu, S. Resveratrol improves alcoholic fatty liver disease by downregulating HIF-1α expression and mitochondrial ROS production. *PLoS ONE* **2017**, *12*, e1834268. [CrossRef]

86. Tang, L.; Yang, F.; Fang, Z.; Hu, C. Resveratrol ameliorates alcoholic fatty liver by inducing autophagy. *Am. J. Chin. Med.* **2016**, *44*, 1207–1220. [CrossRef]

87. Hajinejad, M.; Pasbakhsh, P.; Omidi, A.; Mortezaee, K.; Nekoonam, S.; Mahmoudi, R.; Kashani, I.R. Resveratrol pretreatment enhanced homing of SDF-1-preconditioned bone marrow-derived mesenchymal stem cells in a rat model of liver cirrhosis. *J. Cell. Biochem.* **2018**, *119*, 2939–2950. [CrossRef]

88. Bai, Y.; An, R. Resveratrol and sildenafil synergistically improve diabetes-associated erectile dysfunction in streptozotocin-induced diabetic rats. *Life Sci.* **2015**, *135*, 43–48. [CrossRef]

89. Gencoglu, H.; Tuzcu, M.; Hayirli, A.; Sahin, K. Protective effects of resveratrol against streptozotocin-induced diabetes in rats by modulation of visfatin/sirtuin-1 pathway and glucose transporters. *Int. J. Food Sci. Nutr.* **2015**, *66*, 314–320. [CrossRef]

90. Zhao, W.; Li, A.; Feng, X.; Hou, T.; Liu, K.; Liu, B.; Zhang, N. Metformin and resveratrol ameliorate muscle insulin resistance through preventing lipolysis and inflammation in hypoxic adipose tissue. *Cell. Signal.* **2016**, *28*, 1401–1411. [CrossRef]

91. Brawerman, G.M.; Kereliuk, S.M.; Brar, N.; Cole, L.K.; Seshadri, N.; Pereira, T.J.; Xiang, B.; Hunt, K.L.; Fonseca, M.A.; Hatch, G.M.; et al. Maternal resveratrol administration protects against gestational diabetes-induced glucose intolerance and islet dysfunction in the rat offspring. *J. Physiol.* **2019**, *597*, 4175–4192. [CrossRef]

92. Guo, R.; Liu, B.; Wang, K.; Zhou, S.; Li, W.; Xu, Y. Resveratrol ameliorates diabetic vascular inflammation and macrophage infiltration in db/db mice by inhibiting the NF-κB pathway. *Diabetes Vasc. Dis. Res.* **2014**, *11*, 92–102. [CrossRef] [PubMed]

93. Li, A.; Zhang, S.; Li, J.; Liu, K.; Huang, F.; Liu, B. Metformin and resveratrol inhibit Drp1-mediated mitochondrial fission and prevent ER stress-associated NLRP3 inflammasome activation in the adipose tissue of diabetic mice. *Mol. Cell. Endocrinol.* **2016**, *434*, 36–47. [CrossRef] [PubMed]

94. Qiao, Y.; Gao, K.; Wang, Y.; Wang, X.; Cui, B. Resveratrol ameliorates diabetic nephropathy in rats through negative regulation of the p38 MAPK/TGF-beta1 pathway. *Exp. Ther. Med.* **2017**, *13*, 3223–3230. [CrossRef] [PubMed]

95. Nishimura, Y.; Sasagawa, S.; Ariyoshi, M.; Ichikawa, S.; Shimada, Y.; Kawaguchi, K.; Kawase, R.; Yamamoto, R.; Uehara, T.; Yanai, T.; et al. Systems pharmacology of adiposity reveals inhibition of EP300 as a common therapeutic mechanism of caloric restriction and resveratrol for obesity. *Front. Pharmcol.* **2015**, *6*, 199. [CrossRef]

96. Wang, B.; Sun, J.; Li, L.; Zheng, J.; Shi, Y.; Le, G. Regulatory effects of resveratrol on glucose metabolism and T-lymphocyte subsets in the development of high-fat diet-induced obesity in C57BL/6 mice. *Food Funct.* **2014**, *5*, 1452–1463. [CrossRef]

97. Chang, C.; Lin, K.; Peng, K.; Day, Y.; Hung, L. Resveratrol exerts anti-obesity effects in high-fat diet obese mice and displays differential dosage effects on cytotoxicity, differentiation, and lipolysis in 3T3-L1 cells. *Endocr. J.* **2016**, *63*, 169–178. [CrossRef]

98. Parmar, A.; Mula, R.V.R.; Azhar, Y.; Shashidharamurthy, R.; Rayalam, S. Resveratrol increases catecholamine synthesis in macrophages: Implications on obesity. *FASEB J.* **2016**, *S1*, 301.

99. Zou, T.; Chen, D.; Yang, Q.; Wang, B.; Zhu, M.; Nathanielsz, P.W.; Du, M. Resveratrol supplementation of high-fat diet-fed pregnant mice promotes brown and beige adipocyte development and prevents obesity in male offspring. *J. Physiol. London* **2017**, *595*, 1547–1562. [CrossRef]

100. Huang, Y.; Zhu, X.; Chen, K.; Lang, H.; Zhang, Y.; Hou, P.; Ran, L.; Zhou, M.; Zheng, J.; Yi, L.; et al. Resveratrol prevents sarcopenic obesity by reversing mitochondrial dysfunction and oxidative stress via the PKA/LKB1/AMPK pathway. *Aging* **2019**, *11*, 2217–2240. [CrossRef]

101. Cheserek, M.J.; Wu, G.; Li, L.; Li, L.; Karangwa, E.; Shi, Y.; Le, G. Cardioprotective effects of lipoic acid, quercetin and resveratrol on oxidative stress related to thyroid hormone alterations in long-term obesity. *J. Nutr. Biochem.* **2016**, *33*, 36–44. [CrossRef]

102. Cui, X.; Jing, X.; Wu, X.; Yan, M. Protective effect of resveratrol on spermatozoa function in male infertility induced by excess weight and obesity. *Mol. Med. Rep.* **2016**, *14*, 4659–4665. [CrossRef] [PubMed]

MDPI

St. Alban-Anlage 66

4052 Basel

Switzerland

Tel. +41 61 683 77 34

Fax +41 61 302 89 18

www.mdpi.com

*Foods* Editorial Office

E-mail: foods@mdpi.com

www.mdpi.com/journal/foods